Webb Society Deep–Sky Observer's Handbook
Volume 1: Double Stars

Webb Society Deep–Sky Observer's Handbook

Volume 1
Double Stars

Compiled by the Webb Society
Editor: Kenneth Glyn Jones, F.R.A.S.

Enslow Publishers
Hillside, New Jersey 07205

1979

To Maurice Duruy
Double–Star Observer *par excellence*

First joint American–British edition, 1979

First published in the U.K. in 1975 as
The Webb Society Observers Handbook.
Volume I, Double Stars

Library of Congress Cataloging in Publication Data

Webb Society.
Webb Society deep–sky observer's handbook.

First published in 1975– under title: The Webb Society observers handbook.
Includes bibliographies.
CONTENTS: v. 1. Double stars.—v. 2. Planetary and gaseous nebulae.
1. Astronomy—Observers' manuals. I. Jones, Kenneth Glyn. II. Title.

QB64.W35 1979 523 78-31260
In the U.S.A.: ISBN 0-89490-027-7 (v. 1)
In the U.K.: ISBN 0 7188 2433 4 (v. 1)

Manufactured in the United States of America
10 9 8 7 6 5 4 3

CONTENTS

General Preface

Named after the Rev. T.W. Webb (1807-1885), an eminent amateur astronomer and author of the classic Celestial Objects for Common Telescopes, the Webb Society exists to encourage the study of double stars and deep-sky objects. It has members in almost every country where amateur astronomy flourishes. It has a number of sections, each under a director with wide experience in the particular field, the main ones being double stars, nebulae and clusters, minor planets, supernova watch, and astro-photography. Publications include a Quarterly Journal containing articles and special features, book reviews, and section reports that cover the society's activities. Membership is open to anyone whose interests are compatible. Application forms and answers to queries are available from Dr. G.S. Whiston, Secretary, Webb Society, 'Chestnuts', 1 Cramhurst Drive, Witley, Surrey, England.

Webb's Celestial Objects for Common Telescopes, first published in 1859, must have been among the most popular books of its kind ever written. Running through six editions by 1917, it still is in print although the text is of more historical than practical interest to the amateur of today. Not only has knowledge of the universe been transformed totally by modern developments, but the present generation of amateur astronomers has telescopes and other equipment that even the professional of Webb's day would have envied.

The aim of the new Webb Society Deep-Sky Observer's Handbook is to provide a series of observer's manuals that do justice to the equipment that is available today and to cover fields that have not been adequately covered by other organizations of amateurs. We have endeavored to make these guides the best of their kind: they are written by experts, some of them professional astronomers, who have had considerable practical experience with the problems and pleasures of the amateur astronomer. The manuals can be used profitably by the beginner, who will find much to stimulate his enthusiasm and imagination. However, they are designed primarily for the more experienced amateur who seeks greater scope for the exercise of his skills.

Each handbook volume is complete with respect to its subject. The reader is given an adequate historical and theoretical basis for a modern understanding of the physical role of the objects covered in the wider context of the universe. He is provided with a thorough exposition of observing methods, including the construction and operation of ancillary equipment such as micrometers and simple spectroscopes. Each volume contains a detailed and comprehensive catalogue of objects for the amateur to locate -- and to observe with an eye made more perceptive by the knowledge he has gained.

We hope that these volumes will enable the reader to extend his abilities, to exploit his telescope to its limit, and to tackle the challenging difficulties of new fields of observation with confidence of success.

Preface
Volume 1: Double Stars

The observations of double and multiple stars has been virtually neglected by amateur astronomers during the last 40 years or more. This is a pity, because here we have a field in which the skilled amateur observer can still accomplish much useful work. However there are signs that double star observing is now undergoing a welcome revival, a state of affairs which the Webb Society can claim to have fostered actively during the last decade.

The key to success in this field lies in maintaining sound observing method and obtaining dependable measurements of the elements of position angle and separation. Experience has shown that these goals are well within the attainment of amateur astronomers, using quite modest aperture telescopes, who are willing to obtain, - or make for themselves - a simple micrometer of good design.

This volume provides the observer with all the information necessary to guide him towards successful double star work, and embodies the results of considerable experience by the Double Star Section of the Webb Society over the years. In the text methods of observing and recording observations are set out in detail, and descriptions are given of micrometer equipment, including notes on the construction of several effective types suitable for the amateur to make himself. In addition we provide a detailed catalogue of 3876 measures of 522 double and multiple stars, compiled by the Webb Society. These reliable measures are derived from previously unpublished observations extending from 1941 to 1975, and should prove a useful guide for all observers who have made double stars their special field, or for those who are attracted to pursue this study for the first time.

Almost the whole of the writing of the text, the compilation of the catalogue and many of the measures it contains have been carried out with great thoroughness by Robert Argyle, who is both a Scientific Officer at the Royal Greenwich Observatory, Herstmonceux, and amateur Director of the Webb Society Double Star Section. All this has been accomplished in whatever spare time he could find between his duties as a Night Assistant on the 98-inch Isaac Newton Telescope and his studies for an M.Sc degree. Except for Chapter 6 (written by E.G.Moore) and Chapter 7 (written mainly by the Editor), Robert Argyle can be considered the'author' of this book.

In organizing the production of this handbook, the editor owes a great debt to the skill and experience of Eddie Moore who, as a professional broadcasting engineer, has been engaged in the design and installation of local radio stations in the U.K., and who has long been a tower of strength in many fields of amateur astronomy.

The Webb Society, in the dedication of this volume acknowledges its gratitude to M. Maurice Duruy of Beaume Mele Observatory, Le Rouret, France, who has contributed so many of his superb observations of Double Stars, and whose measures comprise by far the greater part of the catalogue. M. Duruy is an amateur astronomer of great distinction (a short biography is given in Chapter 7) and the Webb Society is greatly honoured in being allowed to publish these results of his work extending over more than 30 years.

AUTHOR'S ACKNOWLEDGEMENTS

Robert Argyle wishes to express his gratitude to Sir Richard Woolley and Dr E.M.Burbidge for permission to use the two telescopes at Herstmonceux for double star observations, and to Dr A.Hunter, Director of the Royal Greenwich Observatory, for permission to publish the results. He is also very grateful to the late Mr L.S.T.Symms who gave up his spare time to demonstrate the operation of the 28-inch refractor. The benefit of Mr Symms' great experience with this telescope was much appreciated.

M. Maurice Duruy is indebted to M. M.Dufay, late Director of the Observatory of Lyons for being allowed to use the 32.5-cm Coude refractor for his observations.

PART ONE.

INTRODUCTION

The observation of double stars in Great Britain between about 1779 and 1930 was in a healthy state. Sir William Herschel had given it a great start with his famous work on binary stars and observational work was keenly followed up by men such as Sir John Herschel, Sir James South, Dawes, Smyth, Seabroke, Knott, Gledhill and others. The measurement of double stars was undoubtedly helped by the availability of really fine telescopes by such makers as Cooke and Grubb in this country and the Clark family in the United States. The delivery of the 28-inch refractor to the Royal Observatory at Greenwich in 1893 seemed to crown the achievements of the early observers. This telescope gave the Greenwich astronomers the opportunity of making many valuable measures of the very close binaries being constantly found by Burnham, Aitken and Hussey in the United States, particularly with the 36-inch refractor at Lick. This concentration on very close pairs left the Reverend T.E.H. Espin and W. Milburn in England the opportunity of discovering over 3,600 pairs under 10" separation between 1901 - 1935 approximately.

Espin and Milburn seemed to be working on their own however since the BAA Double Star Section had ceased to exist in 1914 through lack of support. It is interesting to note that they used reflectors, being the first observers to do so since the Herschels. They thus dispelled the myth that only refractors were any good for double star measurement. A well known contemporary, the Reverend T.E.R Phillips, contributed long lists of accurate measures until his death in 1941. After that, however, lists of measures in amateur and professional journals in this country have been practically non-existent. It is difficult to explain why the observation of double stars should have become so neglected by amateurs, not only in this country but throughout the world. Visual doubles are more numerous than variable stars, for instance, and yet organisations such as the AAVSO are flourishing with thousands of observations being made each year.

One answer to the mystery may hinge on the fact that double star observations are essentially long-term. Some binaries have periods of such length that even after a lifetime of measures many only show that the position angle has moved by a few degrees. In these cases only the observers of the future can reap the rewards supplied by their predecessors. The author considers that the problem is not so much philosophical as practical, in as much that many observers consider that double star measurement can only be done with telescopes of large aperture and filar micrometers. This is, in fact, far from the truth. Simple home-made micrometers exist which can give results of surprising accuracy even with small telescopes. These instruments, known as the binocular micrometer and the diffraction micrometer, have been extensively used by Maurice Duruy, a member of the Webb Society Double Star Section who has been making double star measures since 1934.

Although at present M. Duruy uses telescopes which are large by amateur standards (16-inch and 24-inch) he spent some time making measures with a 10-inch reflector in Montlhery, near Paris, using both micrometers. When the diffraction micrometer was used, pairs down to 1"·0 could be measured and using the binocular micrometer a limit of 0"·5 was reached. M. Duruy records that on nearly every clear night the seeing with the 10-inch was good enough to allow accurate measures to be made. M. Duruy's two micrometers are described later and both are fairly simple to build, particularly the diffraction micrometer, which can be made in an afternoon. This instrument consists of a coarse objective grating which allows 50% of the incident light through and thus only fairly bright pairs can be measured with small telescopes. This is no great disadvantage, however, since measures are required by the Double Star Section of all the bright pairs. The measures which M. Duruy makes are regularly sent to astronomers at the Observatories of Meudon and Nice in France and Uccle in Belgium and he has been kind enough to forward his measures of Struve pairs (both F.G.W. and O.) for inclusion in the Webb Society's Double Star Catalogue.

The number of measurable pairs available to an observer depends on the aperture of his telescope, (not forgetting the state of the atmosphere!). However it is also the case that the cost of a telescope also depends on the aperture and a prospective observer will find that a 12-inch mirror and flat will generally be in the region of 3 - 4 times more expensive than a 6-inch mirror and flat. It should be borne in mind that mirrors are much cheaper than achromatic lenses of the same aperture, and very often a fairly large mirror and flat can be obtained for the same price as a much smaller, though complete, refractor. However, very serviceable mountings and tubes can be made of wood and this can help reduce the cost of acquiring a telescope. The reflector also has several other advantages over the refractor. It is more compact, leading to smaller observatories, and if a diffraction micrometer is being used, the grating can be comfortably adjusted at the eyepiece, whereas in a refractor some sort of arrangement would be required to rotate the grating from the eyepiece end. In the reflector also, if the grating bar width is made to the appropriate size, the effect of the secondary mirror cell can be masked out very easily. The inherent lack of chromatic aberration (given well-corrected eyepieces) is useful when estimating the colours of double star components.

So far the measurement of position angle and separation has been emphasized, but it is also desirable to have accurate values for the magnitudes of double star components. Many of the fainter pairs in the Lick Index Catalogue (I.D.S.) have been given magnitudes by their discoverers and have not been observed since. There could be some interesting results awaiting the observer keen enough to check over these pairs again.

Introduction

Thus there is ample opportunity for useful observations to be made and anyone reading this Handbook, interested in the observation of double stars is cordially invited to contribute to the Double Star Section programme and to help give these underrated objects the attention they undoubtedly deserve.

1. A BRIEF HISTORY OF DOUBLE STAR OBSERVING.

The systematic observation of double stars began in earnest in the year 1779 when William Herschel began the famous series of observations which finally led to his paper of 1803 in Philosophical Transactions, when the existence of binary stars was proved beyond doubt.

Previous to this, very little work had been done, and it is interesting to note that Huyghens found theta Orionis to be triple in 1656 (the fourth component being found in 1684) and yet it was 1755 before the two components of beta Cygni (Albireo) were discovered. Considering that the Trapezium is the more difficult of the two, this illustrates the neglect which double star observing suffered during the first century and a half of the telescopic era. The first pair to be discovered in a telescope was zeta Ursae Majoris (more accurately Mizar, the brighter of the naked eye pair) by Riccioli in Bologna in 1650. It was followed by the Trapezium and on February 8 1665 Robert Hooke discovered gamma Arietis whilst following Comet Hevelius. In the southern hemisphere alpha Crucis was found in 1685 and alpha Centauri in 1689. Two important binaries, gamma Virginis and Castor, were found in the years 1718 and 1719 respectively, the latter being found by Bradley who made an estimate of the position angle of the pair which later helped to fix the period of revolution. The first real list of double stars appears to have been made by Christian Meyer of Mannheim using an eight foot mural quadrant with a power between 60 and 80. His catalogue, published in 1781, included gamma Andromedae, zeta Cancri, alpha Herculis and beta Cygni.

In 1779 Herschel set out to investigate stellar parallax using accurate measurements of close, very unequal double stars. He assumed that the apparent brightness of a star was directly related to its distance. Thus if in a pair of stars one component appeared much fainter than the other, it was reasonable to suppose that the fainter star was, in reality, much more distant. Hence Herschel intended to measure these pairs from different points in the Earth's orbit in order to ascertain the change in position of one with respect to the other, the motion being due to the parallax of the brighter star. In order to find suitable pairs, he began searching with his telescope and soon published a list of 269 pairs, 42 of which were already known. (Apparently Meyer had already tried to ascertain the motion between unequal pairs but Herschel considered that his (Meyer's) instruments were inadequate to show the close pairs required). Herschel continued to measure numerous pairs and after a while he found that some stars had indeed changed position with respect to their neighbours. However the cause of the change was not due to parallax, which should have shown itself over a period of six months. Herschel found that in some cases the motion of one star with respect to the other was not linear and in a beautiful analysis came to the conclusion that orbital motion was the cause. On June 9th 1803, his paper entitled "Account of the

History of Double Star Observing.

Changes that have happened during the last Twenty-Five Years in the Relative Situation of Double Stars; with an Investigation of the Cause to which they are owing", appeared in Philosophical Transactions. This was the first great step forward in double star astronomy.

In addition to showing that binary stars existed, Herschel introduced systematic methods of observation and discovery which were sadly lacking in his time. His telescopes were by far the largest that had been used on double stars, and his home-made micrometers were good enough to give reliable results on fairly close pairs. Only an observer with Herschel's skill, however, could operate the altitude and azimuth motions of his reflector whilst making micrometric measures and taking notes. His last list of discoveries of 145 pairs was made before 1802 but was not published until 1822.

In 1816 Herschel's son John, together with Sir James South, began to review his double stars (Hh). They remeasured virtually all of the H stars as well as producing a large list of new pairs (Sh).

Two years previously, in 1814, Friedrich Georg Wilhelm Struve in the poorly-equipped observatory at Dorpat decided to compile a general catalogue of all known double stars. Although he had only an eight-foot transit circle and a five-foot telescope by Troughton (power 126), he began to observe the positions, and occasionally the position angles and separations, of double stars. In 1820, he published the catalogue of double star positions, and in 1822 his"Catalogus 795 Stellarum duplicium" appeared. In 1824, a refractor with an object glass by Fraunhofer some 9·6" in diameter was delivered to Dorpat. It was mounted equatorially, and the driving clock was of such quality that objects would stay in the centre of the field at a power of 700. The focal length of the objective was 14 feet and a battery of 10 eye-pieces supplied Struve with magnifications ranging from 86 to 1500. He set to work immediately on his great survey of the heavens from the North Pole to -15° declination for the purpose of discovering new double stars and of the formation of a general catalogue. From 1824 until 1835, Struve and his two assistants devoted themselves to the work and in 1837 the great "Mensurae Micrometricai stellarum duplicium et multiplicium"was published. Altogether this work contained 3,110 different systems, some 2,343 of which were Struve's own discoveries, and each system was measured at least three times. Struve's skill and the quality of the 9·6 inch refractor resulted in this catalogue representing a great step forward both in the total number of pairs listed and in the accuracy of the measures obtained. Struve himself paid tribute to the excellence of his telescope, declaring;- "we may perhaps rank this enormous instrument with the most celebrated of all reflectors, viz Herschel's".

In 1839 the observatory at Poulkova was established and Struve became the first Director. A 15-inch refractor, then the largest of its kind in the world, was delivered and the first programme of work was to re-survey all of the stars in the Northern hemisphere down to magnitude 7·0, - some 17.000 in all. This was completed by December 7, 1842, and in that time 514 new pairs had been found by Otto Struve who had taken over the survey from his father a month after it had commenced. Otto Struve added another 33 discoveries, and the catalogue of 547 pairs, known as the Poulkova stars was published in 1850 (106 of the original 514 being omitted for various reasons).

With the completion of this work there was a lull in discovery for about 20 years. At this time several celebrated amateur observers were making their mark in this country and abroad.

Admiral William Smyth began observing in 1830 with a 5·9-inch refractor on a mounting by Dolland. The object glass by Tulley was considered by Smyth "the finest specimen of that eminent optician's skill". Between 1830 and 1843 he made measures of 680 pairs. His "Cycle of Celestial Objects", which also included extensive notes on other deep sky objects was published in 1844. In 1860, the "Speculum Hartwellianum" containing later measures was published.

The Reverend W. R. Dawes began observing in 1831 at Ormskirk in Lancashire with a 3·8-inch achromat by Dolland. He later obtained a 6-inch Merz refractor followed by a 7½-inch Alvan Clark and an 8¼-inch Alvan Clark which was set up at Haddenham in Buckinghamshire in 1859. Finally in 1865 an 8-inch Cooke was obtained, with which he continued to make measures of double stars. His work, which extended from 1830 to 1868 was of a high standard, and, indeed, he was referred to as "the eagle-eyed" by the Astronomer Royal, Sir George Airy. His catalogue, published in 1867, is enriched by the addition of valuable introductions, notes and lists of measures made by other observers.

The most valuable work in the field of double stars was done by Baron Ercole Dembowski who was one of the most celebrated of observers. He began his observations in Naples in 1852, using a 5-inch refractor without a driving clock or a position-angle circle on his micrometer. By the year 1859, some 2,000 sets of measures of high quality had been secured. Only an observer of consummate skill and the highest dedication could have produced such accurate results with such limited equipment. In the same year, he obtained a 7-inch Merz refractor with circles, micrometer and a good driving clock, and between 1862 and 1878 made a further 21,000 measurements. These included all of the Struve stars except 64 which were too difficult for his telescope. Some 3,000 sets were of the Poulkova stars whilst another 1,700 refer to the catalogue of Burnham and others. After his death in 1881, Dembowski's measures were published in two fine volumes in 1883-1884 and today they are regarded as being as valuable as the "Mensurae Micrometricae" itself.

It can be said that the second great period of discovery was initiated on April 27, 1870 when an unknown American amateur, Sherborne Wesley Burnham found his first double star. He was a stenographer in the U.S. Court in Chicago during the day and had become interested in astronomy, obtaining his first telescope, a 3-inch refractor in 1861. This was followed by a $3\frac{3}{4}$-inch equatorial and finally in 1870 he obtained the 6-inch Alvan Clark refractor which he mounted in a home-made observatory in the backyard of his home in Chicago. Armed with a copy of Webb's "Celestial Objects.." he commenced his observations. By 1873 his first list of 81 new discoveries was published in the Monthly Notices of the Royal Astronomical Society. His article began with the words, "It is believed that the stars ennumerated in the accompanying list have been hitherto unknown as double stars, as they are not found noted in the numerous catalogues and publications relating to this subject." Burnham was possessed of remarkably keen eyesight and some of his discoveries are difficult objects in much larger telescopes. His great ability and skill earned him the key to virtually all of the major American observatories. He made discoveries with a large number of telescopes including the 36-inch refractor at Lick, the 40-inch at Yerkes and the 26-inch at Washington. In a career spanning 42 years he discovered 1,340 pairs, many of which are in rapid orbital motion. Whilst at Lick Observatory, Burnham often used to help and encourage younger onservers. One of the most notable to benefit from his advice was Robert Grant Aitken who arrived at Lick in June 1895.

In 1899 Aitken and Professor W.J. Hussey initiated at Lick a survey of all of the stars in the Bonner Durchmusterung (B.D.) down to magnitude 9·0 and as far south as -22° declination, not known to be double. Hussey left Lick in 1905, but Aitken persisted until 1915, when the survey was complete apart from the area of sky from declinations -14° and -22° and between 1 hour and 13 hours Right Ascension. By agreement with the observers at the Union Observatory in South Africa, Aitken was able to extend this survey down to -18° in the 12 hours mentioned. The result was the discovery of 1,329 pairs by Hussey and 3,105 by Aitken, the great majority being separated by less than 5".

At this time, the most modern General Catalogue was Burnham's (1906) which contained details of 13,665 pairs. The great number of new double stars discovered at the Lick and Union observatories made it necessary for a thorough revision of Burnham's catalogue. Aitken imposed stricter limits on the magnitudes and separations involved, with the result that some pairs were rejected. Aitken's catalogue was finally published in 1932 and is known at the A.D.S. It contains 17,180 pairs, many of which are binary in nature.

In England at the beginning of this century, interest in double stars was still strong. There was a Double Star Section in the BAA but this unfortunately petered out in 1914 due to lack of support. The Reverend T.E.H. Espin had initially been interested in cataloguing red stars, but from 1901 onwards he bagan to publish lists of new double stars which were discovered using a $17\frac{1}{4}$-inch Calver reflector mounted in an observatory at Tow Law in County Durham. The first few pairs were casual discoveries, but later he set out to examine all of the stars in the Bonner Durchmusterung north of +30° for uncatalogued double stars of 10" separation or less.

he added a 24-inch reflector later*, and ably assisted by W. Milburn continued his observations for over 30 years. He published all his discoveries in the RAS Monthly Notices, and his last pair bears the number 2,575. Milburn carried on working and eventually reached a total of 1,051 pairs. It is interesting to note that Espin and Milburn were the only major observers since Herschel to use reflectors.

The delivery of the new 28-inch refractor to Greenwich in 1893 allowed the observers there to make numerous and valuable measures in the Burnham, Hough, Hussey and Aitken catalogues as well as the closer Struve and Poulkova stars. The lists of measures can be found in "Greenwich Observations" which extend from 1893 onwards. The chief observers were T.Lewis, H.Furner, W.Bryant and F.Dyson (later to become Astronomer Royal) - names not unfamiliar to those acquainted with Vol II of Webb's "Celetial Objects". Lewis, in particular,was noted for his great summary of all the measures of the Dorpat catalogue in his book "Measures of the Struve Double Stars". When the 28-inch refractor was moved down to Herstmonceux in 1956, work on double stars was continued by Sir Richard Woolley, L.S.T.Symms, M.P.Candy and other observers until September 1970. Early in 1972, however, the telescope was taken back to Greenwich. It is now housed in a replica of the famous "Onion" dome in which it originally resided. In 1975, on the 300th anniversary of the founding of the Royal Observatory at Greenwich, the telescope was inaugurated by Her Majesty the Queen. The telescope is now being used by both professional and amateur astronomers and the filar and comparison image micrometers are once again in use.

In the Southern hemisphere, the serious task of discovering and cataloguing double stars did not really start until the arrival of John Herschel at the Cape of Good Hope in 1834. Using his father's 20 ft reflector (with several 18¼-inch mirrors) and a 5-inch refractor, he catalogued 2,102 new pairs in four years, publishing the results of these observations in 1847. Herschel continued to contribute double star measures regularly to the RAS and was working on a general catalogue of double stars when he died in 1871 aged 79.

It was not until 1896 that serious double star work in southerly latitudes was resumed when Dr R.T.A. Innes joined the staff of the Royal Observatory at the Cape of Good Hope. Using 7-inch and 18-inch refractors he brought his total of discoveries to 432. In 1899 he produced a First General Catalogue of Double Stars, containing 2,140 entries. In 1903 he became Government Astronomer at the Union Observatory

* The 17-inch and 24-inch telescopes with their mirrors were acquired by David Sinden of Sir Howard Grubb Parsons, who has since re-figured the mirrors. The 24-inch is being housed in a dome at Wylam in Northumberland under the auspices of the University of Newcastle-upon-Tyne. The 17-inch mirror is the main component of an advanced catadioptric telescope owned by Alan Heslop of Fulwell, near Sunderland.

and in 1925, a 26½-inch refractor was installed. When Dr Willem H. van den Bos joined the staff from Leiden in Holland, a survet similar to that of Aitken and Hussey in the northern sky was set in motion. Dr van den Bos carried out the major part of the survey, helped by Dr W.S.Finsen and Dr Innes (until his retirement in 1927). At the end of 1931, Innes had found 1,613 pairs, Finsen 300 and van den Bos more than 2,000. At the nearby Lamont Hussey Observatory at Bloemfontein where a 27-inch refractor had been set up, a co-operative programme with the Union observers produced another 4,712 discoveries.The Director, Professor R.A.Rossiter found 1,961 new pairs, whilst his two assistants M.K.Jessup and H.F.Donner found respectively 1,424 and 1,327 new double stars. A catalogue of "Southern Double Stars" was published by Rossiter in 1955 and contains 8,065 pairs.

Mention must be made also of the work done by Robert Jonckheere over 56 years ending in 1962. Between 1909 and 1914, Jonckheere discovered 1,319 pairs at the University of Lille Observatory, and during the First World War continued his observations at Greenwich, where he added a further 252 pairs using the 28-inch refractor. He later continued his work with the 80cm refractor at Marseilles and by 1962 he had discovered 3,350 pairs.

The last (and so far, largest) catalogue of double stars is the Lick Index Catalogue (I.D.S.) which appeared in 1963. It contains details of some 65,000 pairs of which about 40,000 are thought to be binary.

At the present time double star work is being carried out at several professional observatories throughout the world. Brief details of some of the work done at the more important stations are given below:-

(a) Observatory of Nice

For several years much of the work has been done by Dr Paul Couteau. Most of the double stars were observed with the 50-cm refractor and many of Dr Couteau's discoveries have been made with this telescope. To Dec., 1975, the number of new pairs accredited to this astronomer is 1,250. In June 1969 the 76cm-equatorial came back into service and it too is being used for double star measurement.

(b) Observatory of Paris-Meudon

Until fairly recently the principal observers were Dr Paul Baize (now retired) and Dr Paul Muller, who is now based at the Centre d'Etudes et de Recherches Geodynamiques et Astronomiques at Grasse in the south of France. Dr Muller's discoveries to November 1975 number 565 and he continues to edit the IAU Circulars of Commission 26 which contain new discoveries, new orbits and notes of special interest to double star observers.

Dr Baize was for 40 years associated with the 38-cm equatorial and made many thousands of measures with it. He is an assiduous orbit computer and has published several hundred many of which may be found in the Third Catalogue of Double Star Ephemerides and in the latest orbit catalogue by Finsen and Worley, published in 1970.

(c) United States Naval Observatory (Flagstaff,Washington)

The telescopes in use here are the 12-inch and 26-inch refractors and the 24-inch reflector at Washington and the 40-inch and 61-inch reflectors at Flagstaff, Arizona. Since 1962 tha main observers have been C.Worley (14,100 measures) and R.L.Walker (4,800 measures). Both these observers have carried out observing programmes at observatories such as Lick and Cerro Tololo. In addition a photographic programme is being carried out and since 1962 some 9,800 plates have been obtained.

(d) Observatory of Belgrade

Observations with the 65-cm f/16.2 Zeiss refractor began late in 1951 under the direction of P.M.Djurkovic. In 1974 a complete list of double star measures made between 1951 and 1971 was compiled by G.M.Popovic and published by the Observatory. This volume contains 6,527 measures of 1,641 pairs which were made by P.M.Djurkovic, L.M. Dacic, G.M.Popovic, D.J.Zulevic and D.M.Olevic. In addition more than 100 pairs have been discovered by the Belgrade observers and these finds are regularly published in the Circulars of Commission 26.

(e) Sproul Observatory (Swarthmore,Pennsylvania)

Interest here is maimly oriented towards photographic astrometry of visual and astrometric pairs. Dr P.van de Kamp has been a leading investigator in this field for many years. Also on the staff are Miss S.L.Lippincott, current President of IAU Commission 26 and Dr W.D.Heintz, who in addition to publishing many orbital analyses also observes pairs with the 24-inch refractor.

(f) Poulkovo Observatory

Since 1968, some 2,000 multiple exposure plates of 200 double and triple systems have been obtained with the 26-inch Zeiss refractor. It is hoped to find pairs with orbital motion which have changed by more than 10 degrees since 1910-30. Determination of dynamical and trigonometrical parallaxes will then be carried out.

(g) University of Minnesota

Professor W.J.Luyten has continued with his work into common proper motion pairs from plates taken in the Palomar and Bruce Proper Motion Surveys. He has now discovered about 5,000 pairs in this manner including several hundred pairs which contain a degenerate component and several dozen which consist of two degenerate stars.

2. TYPES OF DOUBLE STARS.

Although in this Handbook we are mainly concerned with the observation of visual double stars, it is interesting to summarise the several different types of stellar pairs which are to be found in the heavens. They can be divided into two main categories.

(A) Those which can be detected as double at the eyepiece of the telescope. There are two main types in this category.

(i) Visual Binaries.

The components of these systems are physically related, rotating about their common centre of gravity in periods ranging from less than 2 years to many centuries. When measuring visual binaries with micrometers, it is the convention of astronomers to regard the brighter star as fixed and to determine the distance (ρ) and position angle (θ) of the fainter star or comes. By plotting ρ and θ the comes is seen to describe an elliptical path with the primary situated in one focus.

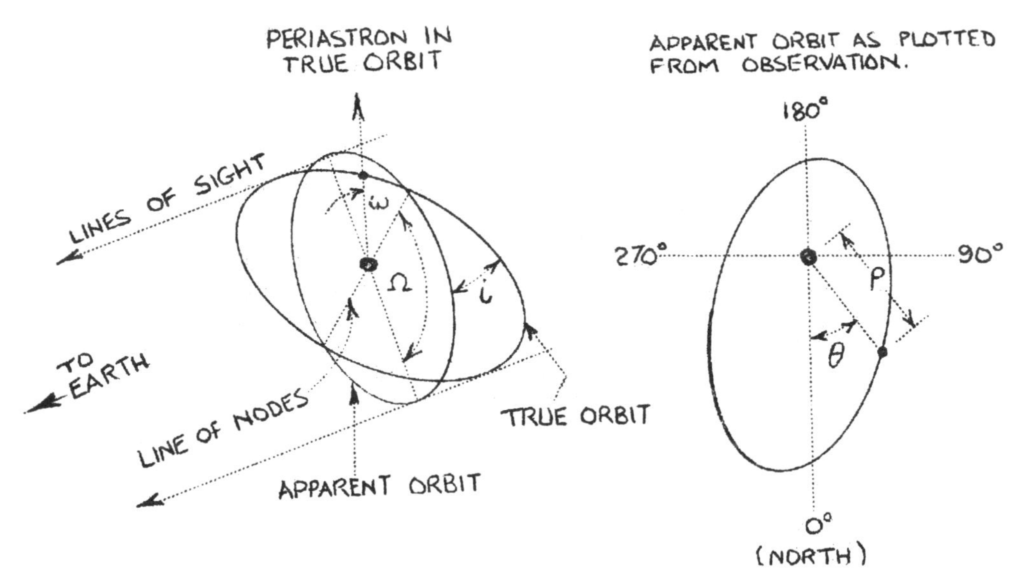

Fig.1. The true and apparent orbits of a double star.

The ellipse is known as the apparent ellipse because it is only a projection of the true ellipse on the plane of the sky. Knowledge of the size and shape of the apparent orbit enables the elements of the true ellipse to be found. There are seven elements required to define the size, shape and orientation in space of the true orbit and the period of revolution of the comes.

Types of Double Stars.

(a) The parameter i is the inclination of the true orbit to the plane of the sky. If $i = 90^\circ$ for instance, the orbit will be edge on to us and the apparent motion of the comes would cause it to oscillate in a straight line. Such a system is presented by alpha or 42 Comae Berenices which is also known as Struve 1728. Here the position angle is fixed at 10° or 190° and the separation reaches a value of 0"·6 at its widest. Orbital elements for edge-on systems are more difficult to compute than usual, since only the measures of separation are of use. It is made doubly difficult if, as in this case, the stars are identical in colour and brightness, since it is very difficult to decide which star is which after occultation occurs.

(b) The nodes are the points where the true and apparent orbits intersect. In the true orbit, ω is the angle between the line of nodes and the direction of periastron - the point at which the comes is closest to the primary.

(c) The position angle of the line of nodes is called Ω - it always lies between 0° and 180°.

(d) The period of revolution in years is P.

(e) The time at which periastron passage occurs is time T. This is the point at which the velocity of the comes is at its maximum.

(f) The eccentricity of the true ellipse is denoted by the letter e. Elliptical orbits have values of e ranging from 0 to 1.

(g) The semi-major axis of the true ellipse in seconds of arc is a.

(h) Sometimes orbit computers quote a value μ or mean motion, which can be found simply from $\mu = 360^\circ/P$.

The orbital elements for a pair, combined with a knowledge of its parallax, will yield a value for the sum of the masses of the individual stars, $m_1 + m_2$. The absolute orbit of each star around the centre of gravity is required before the mass ratio can be found, enabling m_1 and m_2 to be calculated. This is done by measuring each star relative to fixed background stars and correcting for parallax and proper motion.

The best example of a binary star in the northern hemisphere is gamma Virginis (Struve 1670). Its two components of magnitude 3·6 and 3·7 rotate around each other in 172 years. At the moment they are closing up and by the year 2008 they will be only 0"·39 apart and hence visible in large telescopes only.

(ii) Optical Doubles.

In these systems the two stars appear to be close together but they are in no way linked. This is a line-of-sight effect with one star being much further away than the other. The motion of both stars will be rectilinear and uniform if they are not gravitationally connected. Some cases of doubt remain in faint wide pairs where any motion may be small. The proper motion and parallax of such a pair, if measurable, will decide if they are optical or binary in nature. A good example of an optical pair is delta Herculis (Struve 3127) which has two components of magnitudes 3·2 and 8·3. In 1821 F.G.W. Struve measured the position angle as 171°·6 and the separation 27"·8 whilst the I.D.S. quotes a measure of 236° and 8"·9 made in 1958. At the moment motion is quite rapid and the author made a measure of position angle on 1972·67 which was 249°·8.

(B) The second main category of double stars consists of those which are so close that they cannot be detected either visually or with the interferometer.

(i) <u>Spectroscopic Binaries.</u>

It was discovered in 1889 that the spectral lines of zeta Ursae Majoris A (the brighter component of Mizar) were doubled at regular intervals. By comparing the spectrum of the star with a laboratory standard spectrum, it was found that each component of a doubled line varied in step with the other component. This indicated that there were in fact two stars in orbital motion about each other. The magnitude of the shift is directly proportional to the star's orbital velocity in the direction of the Earth.

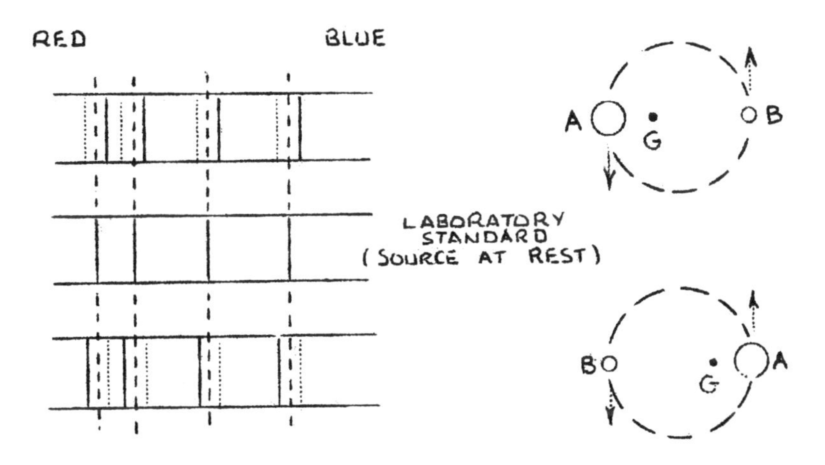

Fig.2. The spectral shift of a spectroscopic binary.

If one component is approaching the Earth the spectral lines will be shifted towards the shorter wavelengths, i.e. towards the blue end of the spectrum. Similarly the other component must be receding and hence its spectral lines will undergo a red shift. When the components of a line are at their greatest separation the stars are moving exactly towards and away from us as illustrated in Fig.2. Similarly when the line is single the two stars are moving in opposite directions at right angles to the line of sight. These are known as double-lined spectroscopic binaries and consist of two fairly equally bright stars. If the difference in magnitude is greater than about 2 the spectrum of the fainter star will not show up, and the pair becomes a single lined system. Instead of plotting position angle and separation, orbit computers plot radial velocity (usually in kilometres per second) against time (which can be in minutes, hours or days). All the normal elements can be obtained from the resulting velocity curve except the orbital inclination i.

A special case occurs when i = 90° and the orbit plane is edge-on to the line of sight. Mutual occultation then occurs and the pair is now known as an eclipsing binary. The consequent light variation serves to give extensive information about the size and shape of the stars and their atmospheres. It must be noted, however, that the pulsating variables such as delta Cephei and RR Lyrae also show these characteristics, i.e. variable light output and variable radial velocity (due to expansion and contraction processes). In general the eclipsing binaries show very distinctive radial velocity curves indicative of orbital motion. However, some cases of doubt remain.

(ii) Astrometric Binaries.

These are apparently single stars which are deduced to be double by virtue of variable proper motion. The classic case is that of Sirius, whose companion was discovered by virtue of its large mass and rapid orbital motion, which produced a well defined periodic wobbling in the motion of Sirius A with respect to background stars. The comes was confirmed visually by Alvan G. Clark in 1862, after being postulated by the great German astronomer F.W. Bessel in 1844.

Strictly, however, astrometric binaries are not visual. A good example is the star zeta Cancri C, whose motion around the close pair was found not to be uniform. Analysis of the measures showed the motion to be due to an invisible companion, the period of rotation being about 17·6 years.

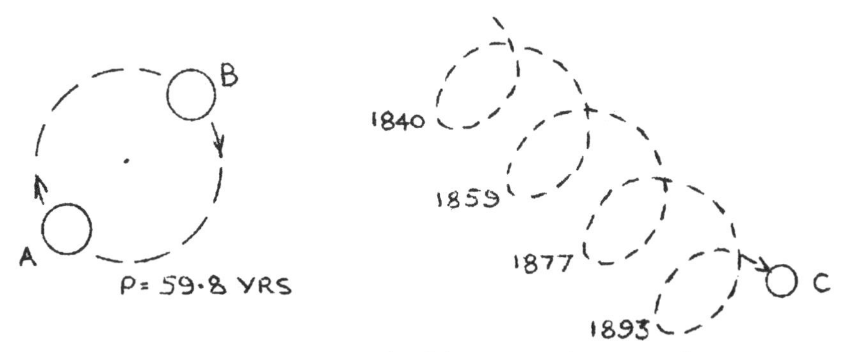

Fig.3. The astrometric binary, zeta Cancri.

In the case of several nearby stars such as 61 Cygni and Barnard's star the perturbation in the proper motion is thought to be due to planetary companions several times more massive than Jupiter.

3. OBSERVING METHOD.

The observation that a star is double in a telescope depends on the following factors:-

(a) The separation of the two components and the aperture and quality of the telescope.

(b) The brightness of the components.

(c) The state of the atmosphere.

(d) The keenness of the observer's vision.

(a) The separation of a double star is usually defined by the angular separation (in seconds of arc) between the centres of the respective Airy disks (this will be explained later).

All stars, with the exception of the Sun, are so distant that it is impossible to see their real disks through any existing telescope. Special techniques, using instruments such as the Michelson Stellar Interferometer, have enabled measurements to be made of the diameters of some of the giant stars such as Betelgeuse, Antares and Mira. Recently, R.Hanbury Brown, using a stellar intensity interferometer at Narrabri in Australia, measured the diameters of 32 stars brighter than magnitude 2.5.

In a good telescope on a night when the atmosphere is steady the appearance of a bright star is that of a small disk. This is not the true disk of the star but the so-called Airy disk formed by diffraction of the incident starlight in the telescope. The diffraction is produced by the aperture through which the starlight is passing and is modified by any additional obstruction in the light path. Thus in a reflector the secondary mirror and its supports will introduce additional diffraction, the spider arms producing "spikes" which extend radially from the stellar disk. Fig 4a shows the idealised image of a star in a "perfect" refracting telescope. In 1834 Sir George Airy found a mathematical relationship which explained the distribution of the light into a central disk surrounded by a system of concentric rings. Airy found that the radius of the first dark ring, r, was given by:

$$r = \frac{1{\cdot}22\,\lambda}{D} \qquad \text{(Fig. 4b)}$$

where λ is the wavelength of the starlight and D the aperture of the telescope in inches. Hence it is apparent that the Airy disk becomes smaller as the diameter of the telescope objective increases.

The Airy formula is sometimes used as a measure of the resolution of a telescope. The Rayleigh criterion states that for a double star to be just resolved, the separation of the two peaks (see Fig 4c)

Observing Method.

should be equal to the radius of the first dark ring. It can also be expressed by:

$$r' = \frac{5''\cdot15}{D(in)} = \frac{13''\cdot1}{D(cm)}$$

assuming that λ = 5700 Å

The value of λ is in fact that of the mean wavelength, since it is not monochromatic but contains the light of many elements, each emitting at different wavelengths. The value of the mean wavelength in fact depends on the spectral class of the star involved and ranges from 5650 Å for a star of spectral class A0 to 5765 Å for one of spectral class K5. In fact the great majority of the components of double stars lie in this region.

Thus the Rayleigh limit depends almost exclusively on telescope aperture and slightly on the wavelength of the incident starlight. It should be noted that magnification has no effect upon resolving power. If a star is double when seen in a low power, then the components will appear to be further apart when higher magnifications are used. If the image of a close double star in the focal plane of the objective is single, then the eyepiece which magnifies the image will always produce a single image irrespective of focal length. The resolution of a double star depends only upon the aperture of the objective.

(b) In the above notes the pair discussed was assumed to consist of two equally bright stars. Although many double stars have fairly equal components, some of the more interesting systems such as Sirius, 85 Pegasi and zeta Herculis, for instance, are very unequal. The Airy disk diameter does not depend upon the brightness of the star and hence the distribution of light in a faint star must be much less marked than in a bright star. Fig 4d shows the situation where a faint star is close to a bright primary - at certain distances the comes is obstructed by the diffraction rings of the primary (if it is bright) and these would probably render the fainter star invisible. This problem occurs in pairs such as delta Cygni (magnitudes 3·0 and 7·5, separation 2"·2 - 1972). Sirius presents a slightly different problem since the separation of the comes Sirius B can exceed 11" of arc (at present, 1972, the pair is approaching its widest separation and thus should be examined carefully on a good night). The magnitude of Sirius B is 8·4 and thus is 10 magnitudes or 10,000 times fainter than the brilliant primary and difficult to see in the glare of the great star. In order to see it at least a 15-cm reflector is required. It is sometimes useful to use a hexagonal diaphragm which fits over the end of the telescope tube, (Fig 4f). This modifies the diffraction pattern and produces a six-pointed star effect which has the result of "diverting" light forming the stellar disk into the spikes thus reducing the size and revealing any close faint stars (Fig 4g). If the diaphragm is arranged to rotate a faint star can be seen between two adjacent spikes where the sky is comparatively dark. A regular hexagon will produce a light loss of only some 17% which is only a fraction of a magnitude.

Observing Method.

A pair of equally faint stars does not conform to the Rayleigh limit because there is not enough light available to detect the small contrast between the stars and the gap between them (Fig 4e). The Rayleigh limit, whilst based on theory, does give a value for the limiting separation, r, which is slightly larger than that found for telescopes in practice. An empirical limit, known as Dawes's Limit after the Reverend W.R. Dawes who first determined it, gave the value:

$$r'' = \frac{4''\cdot 56}{D(in)} = \frac{11''\cdot 58}{D(cm)}$$

This was based on the observed limit of resolution by a number of observers using a range of apertures. It is approximately valid only for a pair of 6th magnitude stars in a small telescope, and does not take into account the cases of unequal or faint pairs. Thomas Lewis and Robert Grant Aitken both determined extensions of the Dawes's Limit to cover these situations. To quote some examples; Lewis, using results of 31 observers with 24 different apertures, found for a pair of mean magnitudes 5·7 and 6·4 that the limit was 12"·3/D(cm), whilst Aitken obtained 10"·9/D(cm) for a pair of 6·9 and 7·1 magnitude stars. For faint pairs, Lewis found 21"·6/D(cm) for a pair of magnitudes 8·5 and 9·1 whilst Aitken got 15"·5/D(cm) for stars of magnitudes 8·8 and 9·0.

(c) The two foregoing sections consider the rather unlikely situation of the atmosphere being perfectly steady. In reality, the images obtained in a telescope rarely, if ever, resemble those shown in the previous sections.

The primary effect which the atmosphere has on star images is to swell them out into disks several times larger than the diameter of the Airy disk. This is known as the seeing, and is often quoted as being the apparent diameter of the star image in seconds of arc. Other effects are often observed; the image may appear to vibrate randomly (known as wandering) or may occasionally swell and contract (boiling). In addition, near the horizon atmospheric dispersion distorts the images into short spectra, and it hardly needs stating that if a close or difficult pair is being examined it should be done when the pair is highest in the sky. It is thus useful to have a list of pairs to be observed which contains a number of close and difficult pairs, so that the observer can be prepared to take full advantage of a night when the seeing is particularly good. It can be a source of great satisfaction when a close binary is seen for the first time after periastron has occured, or a particularly close elusive comes is seen after many attempts on inferior nights.

(d) Another variable concerned with the practical side of double star observing is the keenness of the observer's vision. It is undoubtedly true that some of the great observers such as Burnham, Aitken, Dawes and Dembowski were possessed of remarkably keen eyesight, and this coupled with a lifetimes experience in examining stellar disks, enabled them to detect very small elongations and faint close companions to bright stars. However, the amateur observer should not be unduly put off by this knowledge. The author who cannot claim to have keen eyesight has found that Dawes's Limit can usually be reached on most of the telescopes he has used. However, the concern here is not with very close and rapid binaries, but those double stars comfortably within the reach of amateur instruments.

Having discussed some of the observing problems raised in this particular sphere of observation, it remains to decide what type of observation an amateur observer can carry out. The field can be divided into three distinct sections.

(a) Photographic.

Photography of double stars is somewhat of an unknown quantity in amateur circles and is very rarely done in this country. Later on in this handbook E.G.Moore has devoted a section to this particular subject and anyone interested in trying double star photography will certainly find it a great challenge.

(b) Visual.

This category is mainly applicable to small telescopes (below about 15-cm aperture). In making estimates of magnitude and field drawings the large field is a definite advantage. Most observers of double stars will have done this sort of work at one time or another, and it is very useful for giving the observer a general introduction to the subject. The author observed several hundred pairs, making notes of colours, magnitudes and estimated position angles and separations before making micrometric measures. All of these pairs were found using a copy of Norton's and "Celestial Objects.." Vol II and an eyepiece giving a fairly large field - about 1° in this case. Several other eyepieces of higher power were available for examining the pair more closely after identification had been made. Eyepieces should cover a range of magnifications from about 3A to 25A where A is the aperture in cm. As the telescope aperture increases the upper magnification which the telescope will stand gradually decreases and for the large refractors the value drops to 20A or less.

It is very useful to determine the diameter of each eyepiece field in minutes of arc - this can be done easily by timing the passage of an equatorial star across the field of view. Knowledge of eyepiece field helps to find faint pairs by offsetting, and the position of unidentified pairs for future reference to be noted. It is important that the eyepieces be of good quality if the OG or mirror is excellent so that the best performance can be extracted from the telescope as a whole. Good eyepieces can last a lifetime and of course can be used when larger optics are acquired.

(c) Micrometric.

For those observers with apertures greater than about 15-cm the addition of a micrometer to their equipment can be considered.

Observing Method.

Unfortunately, filar micrometers are rare (and expensive) items in this country and they are not in production by any of the more well-known astronomical instrument makers. One firm in the U.S.A. supplies them but at a cost of about £70 and this makes them prohibitively expensive. The B.A.A. does possess a number of these instruments and those who are B.A.A. members should keep a watch on the list of "Instruments on Loan" should they become available.

However, whilst filars are very useful, they are not necessary in order to make measures of position angle and separation. For brighter pairs the diffraction micrometer, described later, is a very simple alternative and is easy to construct. In the author's opinion the best instrument is the comparison image micrometer, which is more accurate than the filar for measuring separations, but a certain amount of engineering knowledge is required to build one. Most of the optics can be bought in war surplus stores, but the Wollaston prism may be difficult to obtain and rather expensive. It should be noted that the reflector design (Fig 5) does not impede the incoming light in any way and no field illumination is required. The absence of the quarter silvered cube (in the refractor design) makes fainter pairs more easily measurable.

The measurement of binaries is one of the most fascinating and exacting pursuits in observational astronomy. There are a number of bright pairs whose orbital motions are such that changes can be detected in periods of a year or less and which are available to small telescopes. However there are also a great number of wider pairs which have not been measured for some considerable time. The acquisition of a larger telescope provides the observer with a corresponding increase of observational material available and at present even the 12-inch reflector is not an uncommon size amongst amateurs.

Many of the measures in the catalogue were made with 6-inch and 10-inch telescopes and given a number of keen observers, there is no real reason why some useful work may not be done.

Observing Method.

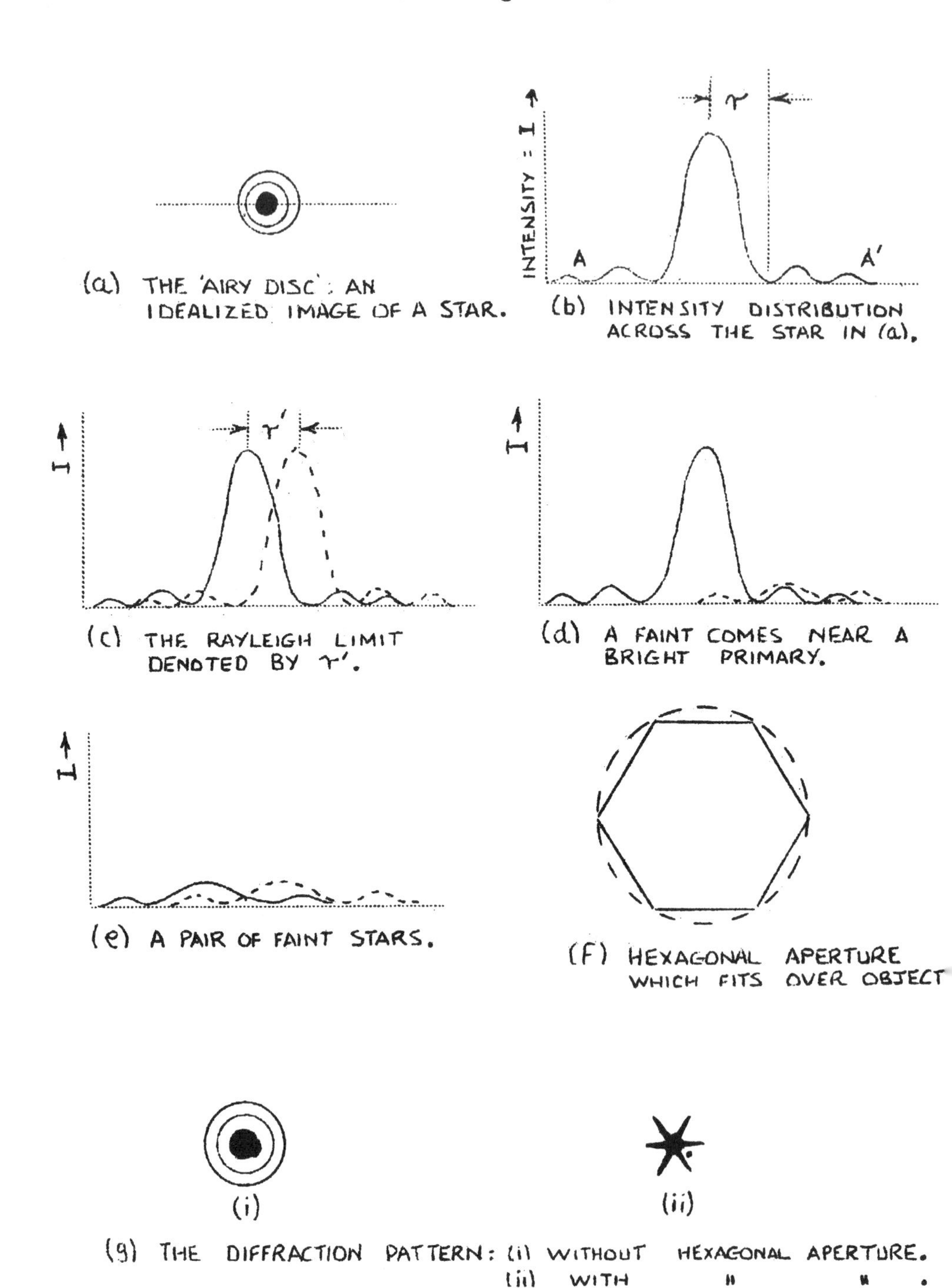

(a) THE 'AIRY DISC': AN IDEALIZED IMAGE OF A STAR.

(b) INTENSITY DISTRIBUTION ACROSS THE STAR IN (a).

(c) THE RAYLEIGH LIMIT DENOTED BY r'.

(d) A FAINT COMES NEAR A BRIGHT PRIMARY.

(e) A PAIR OF FAINT STARS.

(F) HEXAGONAL APERTURE WHICH FITS OVER OBJECT

(g) THE DIFFRACTION PATTERN: (i) WITHOUT HEXAGONAL APERTURE.
(ii) WITH " " .

FIG 4

Observing Method.

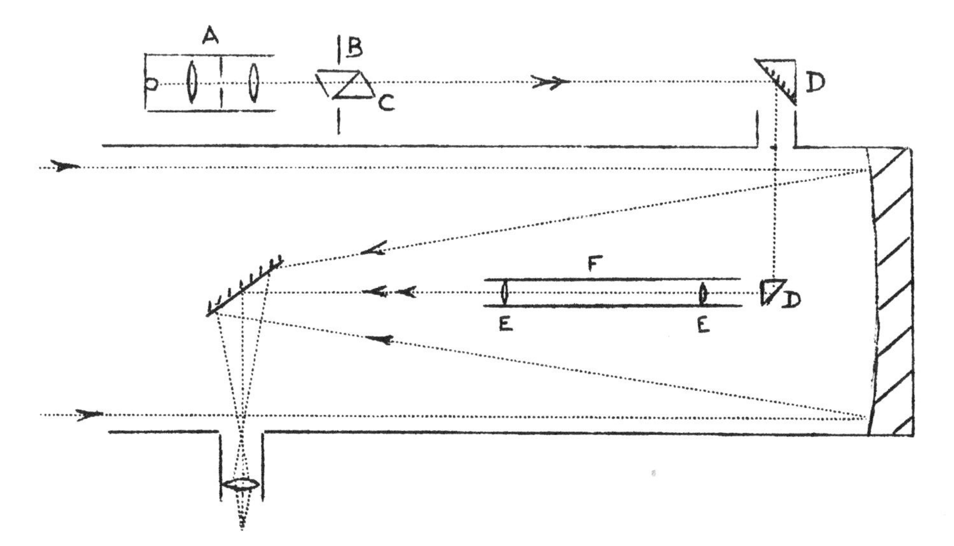

FIG 5. COMPARISON IMAGE MICROMETER FOR REFLECTOR.

A. PINHOLE TO PRODUCE INITIAL IMAGE.
B. POSITION CIRCLE TO MEASURE P.A.
C. WOLLASTON PRISM TO PRODUCE DOUBLE IMAGE.
D. PRISMS OR OPTICAL FLATS.
E. SUPPLEMENTARY LENSES.
F. TUBE TO HOLD SUPPLEMENTARY LENSES ETC.

Further details of this instrument are given on page 36.

4. RECORDING OBSERVATIONS

(a) LOCATION.

The pairs to be observed may be found (if no setting circles are available) by two main methods.

(i) Differentiation from a Known Star.

To employ this method it is necessary to know the diameter in minutes of arc of the field of view of the eyepiece being used. To evaluate this quantity, the telescope should be centred on an equatorial star such as delta Orionis, zeta Virginis or zeta Aquarii. With the telescope stationary, time the passage of the star across the centre of the field from edge to edge - if this time is in minutes then the diameter of the field in minutes of arc can be found by multiplying by 15.

(ii) Sweeping for a Pair.

This is a less certain method than (i) and should be used if the RA and Declination co-ordinates do not correspond to the epoch of the star atlas available.

(b) POSITION.

If a double star is found during observation and cannot be identified, its position in terms of RA and Declination can be obtained by differentiating from a known star. After locating it, the position angle, separation, colours and magnitudes of the pair should be estimated.

Never neglect a pair because it cannot be identified. By giving as much information as possible it should be possible to identify it in the Lick "Index Catalogue of Double Stars".

(c) SEPARATION.

The diameter of the eyepiece field is again useful in estimating separation. If a high power is used then the separation of a wide pair becomes an estimable fraction of the total field diameter, and it becomes quite easy to make a direct estimate of separation. Observers should always make direct estimates of the separation of a pair when possible. If a micrometer is being used, four measures of separation should be made and the mean value corrected to the first decimal place.

Double stars can be classified into five sections in terms of separation:

1. Very Close.	0"·5	-	2"·0
2. Close.	2" +	-	5"·0
3. Standard.	5" +	-	10"·0
4. Wide.	10" +	-	30"·0
5. Open.	30" + and above.		

Recording Observations.

Observers should determine whether the stars in a close pair are single, elongated or divided. If single the fact should be recorded.

(d) BRIGHTNESS.

Observers should always estimate the brightness of both components of a pair to 0·1 magnitude if possible. If the pair is very close then it will be easier to estimate the difference in magnitude of the components and then to use a low power to estimate the brightness of the single image. From these two values the brightness of each star can be determined.

If C is the combined magnitude and A and B are the magnitudes of the brighter and fainter components, and d is the estimated difference in magnitude of the two components, then:-

$$A = C + x.$$
$$B = A + d.$$

Table 1, below, gives the value of x for a range of values of d, and then, since C is known, the values of A and B can be found.

TABLE 1.

d	x	d	x
0·0	0·75	1·2	0·31
0·1	0·70	1·4	0·26
0·2	0·66	1·6	0·22
0·3	0·60	1·8	0·19
0·4	0·57	2·0	0·15
0·5	0·53	2·5	0·10
0·6	0·49	3·0	0·06
0·7	0·46	3·5	0·04
0·8	0·42	4·0	0·03
0·9	0·39	4·5	0·02
1·0	0·36	5·0	0·01

e.g. The pair STT174 was observed to have a combined magnitude of 6·2 and a difference of magnitude of 1·5,

Thus approximately; $A = C + x$ and $B = A + d$, where $C = 6{\cdot}2$ and $d = 1{\cdot}5$, (whence $x = 0{\cdot}24$ from Table 1).

$$A = 6{\cdot}2 + 0{\cdot}24 = 6{\cdot}44$$
$$B = 6{\cdot}44 + 1{\cdot}5 = 7{\cdot}94$$

and correcting to the first decimal place gives 6·4 and 7·9 as the estimated magnitudes of the pair.

(e) COLOURS.

Table 2 (by Hagen) provides for most of the hues which observers are likely to meet during examination of double stars. If a colour which is not in the table, such as purple or green etc. is found, then this should be written out in full in the column when the observer makes out his report. Otherwise the colours can be converted into two numbers which saves space when the results are being entered into the card index.

TABLE 2.
Colours of Double Stars (after Hagen).

-3.	Pure blue.
-2.	Pale blue.
-1.	Blue white.
0.	Pure white.
1.	Yellowish white.
2.	Pale yellow.
3.	Pure yellow.
4.	Orange yellow.
5.	Yellow orange.
6.	Pure orange.
7.	Reddish orange
8.	Orangey red.
9.	Red, slight orange tint.
10.	Pure red.

(f) POSITION ANGLE.

This is estimated by allowing the pair to drift across the field and estimating the angle between a line joining the stars and the direction of drift. The position angle always lies in the direction from the brighter to the fainter star. Fig 6 shows the arrangement in an astronomical (inverting) telescope.

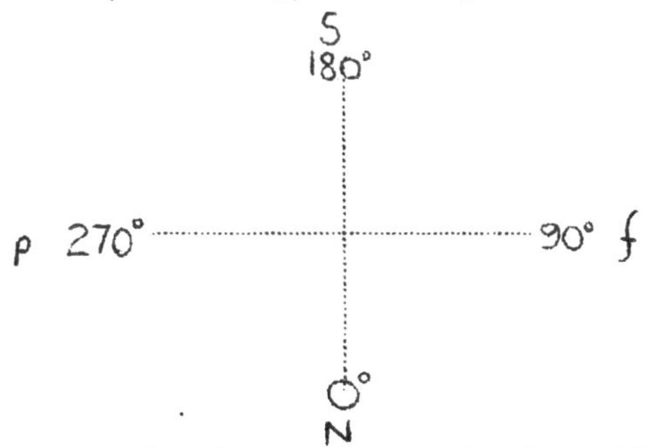

Fig.6. The measurement of position angle.

With practice estimates of position angle can be made usually to within 10°. If a micrometer is used it will usually suffice to make three independant determinations of position angle. The mean figure should be quoted to the nearest 0°·1.

(g) DRAWINGS.

It is not considered necessary to draw fields of identical diameter - it is up to the observer to include features which he may consider interesting. It is always necessary to show the north-south line and to include the scale of the drawing in minutes of arc. Other data included should be: size and type of telescope, magnification employed and brief notes on the sky conditions.

The Antoniadi scale of seeing from I to V can be used to describe the steadiness of the atmosphere, i.e.:-

I. Perfect seeing without a quiver.
II. Slight undulations, with moments of calm lasting several seconds.
III. Moderate seeing with larger tremors.
IV. Poor seeing, with constant troublesome undulations.
V. Very bad seeing.

In order to standardise all field drawings the following symbols have been adopted to represent field stars and double star components.

TABLE 3.
Symbols for star magnitudes.

Symbol		
✦	MAGNITUDE	5
▲	"	6
●	"	7
●	"	8
•	"	9
·	"	10 & BELOW

A specimen observation using the above notes would be:-

1971 July 16 | 10-inch spec. | Seeing: II

Pair: epsilon Sagittae (Otto Struve 185 Appendix)

Separation: 100" ± (Open)

Magnitudes: 5.5,6.5

Colours: +2,-3.

Position angle: 80°

Eyepiece: x80 (Field 65').

A standard reporting form for recording observations, as shown opposite is available to Webb Society Double Star Section observers on application to the Director.

ADS:

Recording Observations.

WEBB SOCIETY DOUBLE STAR SECTION. OBSERVING FORM.

PAIR: CATALOGUE (if any):

RA: DECL: EPOCH:

RA:* DECL:* EPOCH 2000 (Leave * blank)

DATE: OBSERVER:

APERTURE: (State if OG or Spec): MAGNIFICATION(S):

SEEING: MICROMETER (if any):

TRANSPARENCY:

FIELD DRAWING.

N

Scale:**

Magnitudes:

5 6 7 8 9 10 & BELOW

NOTES.

MAGS (Est):

COLOURS:

P.A. (Est):

DIST (Est):

P.A. (Meas):

No. Readings:

DIST (Meas):

No. Readings:

OTHER COMPONENTS: (Give details as above):

** The observer should indicate the scale of the field drawing by means of a line, typically thus: 20'

5. MICROMETERS FOR DOUBLE STAR MEASUREMENT.

In this chapter it is proposed to describe the various instruments which can be used to measure the position angle and separation of double stars. It covers four basic classes of micrometers which will be described as follows.

(a) FILAR MICROMETERS.

The filar micrometer, which is probably the most well known and widely used instrument for this purpose, was first described by Auzout in 1667 in a form which has not changed radically to this day.

Basically the instrument consists of a spider thread stretched across a frame which is itself mounted in a box and free to move to and fro within it. A second thread is fixed across the box parallel to the movable thread and will be referred to as the fixed thread. It is displaced slightly along the optical axis of the telescope to allow free movement of the movable thread. The frame containing the movable thread is controlled by a micrometer screw of fine pitch, (usually graduated in one hundredths of a revolution). A second screw known as the box screw controls the movement of the whole box, hence both threads may be moved together while remaining equidistant. A third thread, also fixed across the box, is inserted perpendicular to the movable and fixed threads and is used to measure position angle. The micrometer is generally able to rotate through 180° in the plane perpendicular to the optical axis, although some instruments have frames which rotate through 90°. The P.A. can be read off from a circle graduated in intervals of 1° or 0°·5, generally by means of two verniers, and illumination of the threads is obtained by a small bulb in the body of the micrometer which is connected to a low voltage circuit with a potentiometer to allow the level of illumination to be varied.

Before a double star can be measured, it is essential to determine the screw value of the movable screw with great care and accuracy and also the zero reading of the position circle P_0. If a fairly bright equatorial star is allowed to drift across the field of view with the driving clock stopped, then by turning the position angle thread until the star drifts exactly along it, the angle on the circle corresponds to $P_0 + 90°$. If the angle on the circle is, for example, 89°·2, then 0°·8 must be added to each measure of P.A. made, and this value must be checked regularly. Thus, in this case P_0 corresponds to 359°·2.

There are two main methods for finding R, the screw value of the micrometer or movable screw.

(i) By measurement of the difference in declination of two stars.

The Pleiades are well suited for this purpose - containing stars of similar magnitude and R.A. and whose declinations are accurately known. Having determined P_0, the micrometer is then arranged so that the fixed and movable threads are both horizontal (i.e., they move in declination only).

Micrometers for Double Star Measurement.

Depending on the micrometer the difference in declination between suitable members is about 50 - 100 turns of the screw and this necessitates the use of stepping by means of intermediate stars. The measures should include a reading from the north star to the south star and back again on several separate nights. Precession, refraction and proper motion should all be taken into account.

(ii) Using the transit of circumpolar stars.

With the circle reading $P_o + 90^\circ$, the star to be used is observed with a low power just before upper or lower culmination. The movable wire is placed just preceding the star as it enters the field, with the driving clock stopped. Using the sidereal timepiece, the transit time of the star on the thread is noted. The thread is advanced one revolution, (or a suitable fraction of a revolution), and the transit time is again noted. This is repeated 30 - 40 times both before and after culmination.

Full details on both of these methods can be found in Aitken's book "The Binary Stars" which is essential reading to all serious observers.

The telescope can now be turned on the double star in question, but before measures are made it is useful to estimate the d_m (difference in magnitude of the two stars and the quadrant in which the fainter lies if they are not equally bright.

The first measurement is that of position angle. The P.A. thread is turned until it appears to bisect the images of both stars simultaneously. The reading is noted on the circle and several more determinations made, remembering to add the correction (in our case $0^\circ{\cdot}8$) to the mean P.A. at the end.

To measure separation, the micrometer is first of all turned through $\theta + 90^\circ$ where θ is the mean P.A. <u>without</u> correction.

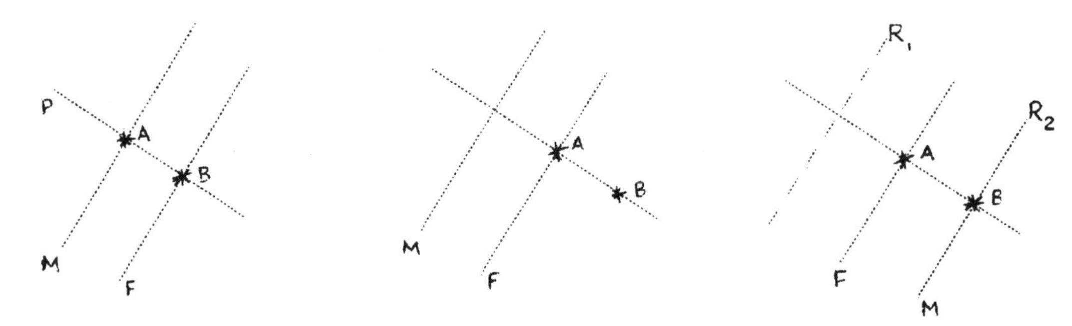

Fig.7. Filar micrometer. Arrangement of threads for measurement of separation.

Fig. 7 shows the positions of the position angle thread (P) and the movable and fixed threads (M and F respectively) during the measurement of separation.

Firstly the micrometer screw and box screws are adjusted until the movable thread bisects the image of the brighter star and the fixed thread bisects the fainter star. The box screw is then adjusted until the fixed thread now bisects the brighter star A. The reading on the micrometer or movable screw (R_1) is noted. Keeping the fixed thread bisecting star A, the movable thread is adjusted until it bisects star B, the new reading (R_2) being noted. The value $R_2 - R_1$ gives twice the number of revolutions required to move between the two components A and B, and knowing R, this distance can be converted into seconds of arc.

The main advantages of the filar are its accuracy and range of application, whilst its drawbacks include:

(a) The illumination of the threads or the field will render very faint comites invisible.

(b) The closeness of the pair may necessitate the use of a magnification which is too high for the state of the atmosphere, resulting in ill-defined and moving images, with reduced accuracy in the measures.

(c) The movable and fixed threads are not parfocal and refocussing is necessary.

Another form of filar micrometer is the bi-filar which has two movable threads and two fixed threads and which is used extensively in cometary photography as a guiding eyepiece.

Jonckheere's skew thread micrometer (1950) consists of two threads on a movable frame which make an acute angle with the axis of the screw and a fixed thread which is oriented so as to bisect the obtuse angle formed at the intersection of the two movable threads.

For pairs too wide to measure with the crossed threads another thread, mounted on the same arrangement of carriage but parallel to the fixed thread can be added. This has a different screw value which is determined as for the filar micrometer.

The distance actually measured is several times greater than the distance between the two stars which form one side of the triangle and thus the screw value is correspondingly smaller - Jonckheere's value being 1"·078 per revolution. Another advantage is that since the movable threads are fixed with respect to each other, they can be stretched in the same plane. The fixed thread is used to measure position angle.

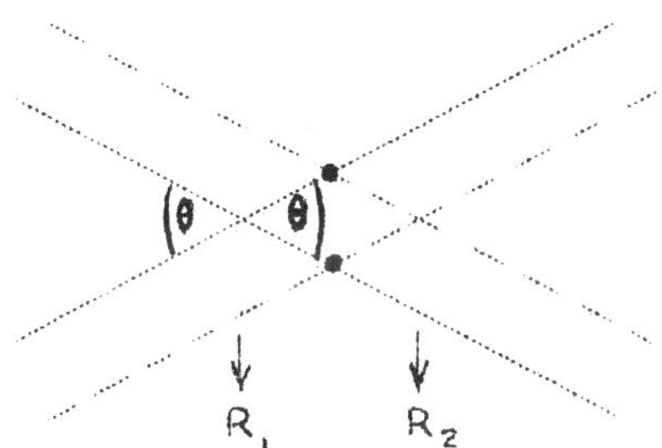

Fig. 8. Jonckheere's micrometer. Method of measuring separation.

Micrometers for Double Star Measurement.

The separation of the pair (ρ) depends upon R_1, R_2, and the angle θ between the fixed threads. This angle can be determined by rotating the micrometer and allowing an equatorial star to drift along each wire in turn. This proceedure should be carried out several times with great care. The separation of the pair is then given by:

$$\rho = (R_2 - R_1)\tan(\theta/2)$$

The advantages of this micrometer are:

(a) Both fixed threads can be placed in the same focal plane.

(b) Only one screw is required for the measurement of separation.

(c) There can be a very small screw constant.

An interesting micrometer developed by Bigourdan (1895) has two glass points attached to fixed and movable frames of the micrometer. The two glass needles are used to "point" at the two components of the double star and measurements are made in the same way as with the filar. However, a thread is still needed for P.A. measurement.

One group of micrometers has no moving parts. The movable and fixed threads are replaced with a graticule at the focal plane. The graticule may have a variety of forms - lozenge, circle, two concentric circles (ring micrometer). The relative co-ordinates of two stars are obtained by noting the instants of contacts as the stars drift over the graticule. The accuracy of such observations is, however, much inferior to those obtained with a filar micrometer.

(b) IMAGE MICROMETERS.

The two main instruments in this category are the double image and the comparison image micrometers. A third instrument, the diffraction micrometer will be discussed in detail later in this chapter.

The most familiar instrument in this group is the double image micrometer which began life in the form of a divided object glass known as a heliometer. However, this arrangement produced aberrations which prevented the use of high powers for the measurement of close double stars. A much easier and more useful method was to place a divided lens just in front of the eyepiece. Thus by displacing one half of the lens with respect to the other along the axis of division, two images of a single star are obtained. By displacing in the opposite direction, the two images appear to merge and open out again. When viewing a double star, therefore, four images are produced and by moving the half-lens (by means of a micrometer screw with a milled head) so that certain combinations of the four images are obtained, the distance between the pair of stars can be found, knowing the screw constant. To make a measure the micrometer has to be rotated until all four images form a straight line, or a square.

For example, let ⦶ and ⦶ represent one pair of images of the double star to be measured and ⊖ and ⊖ represent the other pair of images. In the "zero" position of the half-lens, they co-incide and now suppose that the micrometer screw is rotated until the images appear as in Fig. 9(a). If the distances between all the stars are thought equal then the shift of the micrometer has been D/2, where D is the distance of the double.

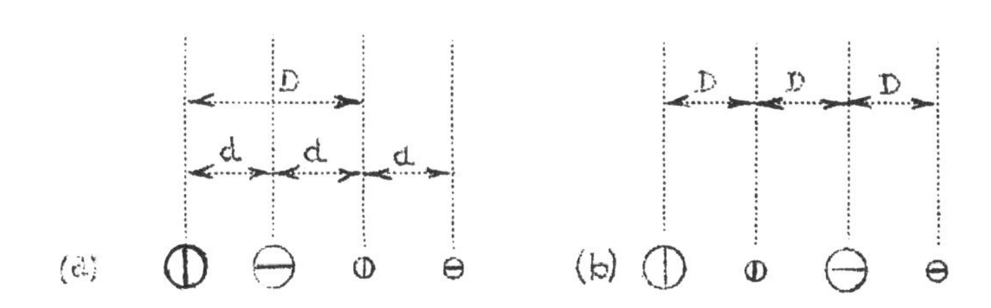

Fig. 9. Image micrometer; Method of measuring separation.

This combination is now obtained on the other side of "zero" and hence the difference in the two micrometer readings equals D, and the zero reading is eliminated. Similarly in Fig. 9(b) the total shift is in fact 4D, this method being preferred for close pairs.

The micrometer is also fitted with a graduated position angle circle and the angle at which co-linearity occurs corresponds to the position angle - remembering to take careful note of the quadrant in which the fainter star lies.

Later instruments such as those used by P. Muller use birefringent, or double refracting prisms, in place of the split lens. In these prisms, which are made of two prisms of quartz or calcite mounted with their crystal axes perpendicular, an incident light ray is split up into two rays known as the ordinary (o) and extraordinary (e) rays. Dr Muller developed this prism (which is now named after him) in 1937. When the prism is moved perpendicularly to the optical axis of the telescope a variation in the separation of the o and e rays is produced, (Fig. 10).

Measures of double stars are made in a similar manner to that of the split lens micrometer. One advantage of the Muller prism is that the o and e rays are polarised and by fitting a thin polarising sheet (which acts as an analyser) the difference in magnitude between the two components can be measured. According to Dr Muller the average mean error for observations made in one night is 0·04 magnitude. Pairs as close as 0"·45 have been measured in this way with the 60-cm at the Pic-du-Midi observatory.

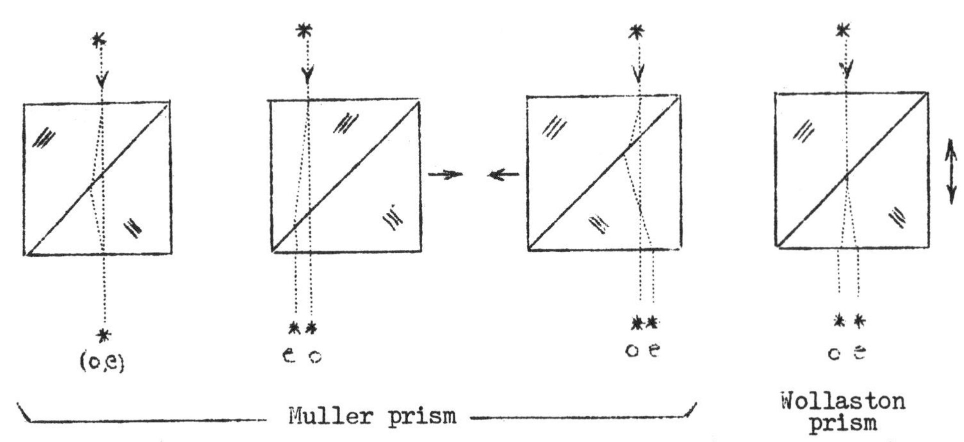

Fig. 10. Birefringent Prism Micrometer (after Muller).

The comparison image micrometer (CIM) which was introduced by F. J. Hargreaves in 1931 produced an artificial pair of stars. A modified instrument was produced by Hargreaves in 1932, and was then adapted for use with the Greenwich 28-inch refractor by L. S. T. Symms and C. R. Davidson.

In this instrument, an illuminated pinhole is imaged onto a Wollaston prism by means of a system of biconvex lenses. The Wollaston prism (Fig. 10) then splits the image into two components which are polarised. The separation of the two images depends upon the motion of the prism towards or away from the image of the pinhole. A centimetre scale is used to measure this movement, which is related to the separation of the artificial stars, which must now be imaged in the eyepiece in sharp focus, alongside the images of the pair to be measured. This is achieved by projecting the images up the outside of the telescope tube to an elliptical flat which reflects them onto a reflecting prism side-by-side with the lamp housing. They are reflected into the telescope tube and fall upon a small object glass of 11½-inch focus, and via a quarter silvered cube into the field of the eyepiece in sharp focus. The Wollaston prism is mounted on a graduated P.A. circle and is free to rotate to enable P.A. measurements to be made. A pair of crossed Nicol prisms, also rotatable, can be adjusted to make either of the artificial stars as bright or as faint as required to match the pair being measured. Finally a blue filter adjusts the colour temperature of the lamp's tungsten filament (2,800°K) to one of 5500°K which is the colour temperature of an average star as seen in the 28-inch.

A movement of 1 cm in the lamp housing is equivalent to a change of separation of 0"·8 of the artificial stars in the eyepiece. The CIM is capable of measuring pairs whose distance ranges from about 4"·5 to less than 0"·2 and it is extremely easy to use. A filar micrometer also fitted to the 28-inch is preferred for P.A. measurement.

Micrometers for Double Star Measurement.

In general, image micrometers have three main advantages over the filar micrometer, these being:

(a) No field illumination is necessary.

(b) There is no masking of the star images.

(c) On poor nights it is easier to compare two pairs of stars than to try to bisect two moving images.

On the debit side, the measurement of position angle tends to produce rather less accurate results than can be obtained with the filar.

(c) THE BINOCULAR MICROMETER

This micrometer was devised by M.V.Duruy in the mid-1930's. It consists basically of two eyepieces, one of which forms the image of a double star from the telescope, the other images an artificial double star produced by one of the methods described later in this chapter. When viewing both pairs of stars simultaneously it is possible to detect very small differences in the observed separations. Knowing the focal length of each eyepiece, the true separation of the artificial pair in seconds of arc can be calculated, knowing the actual separation in say mm.

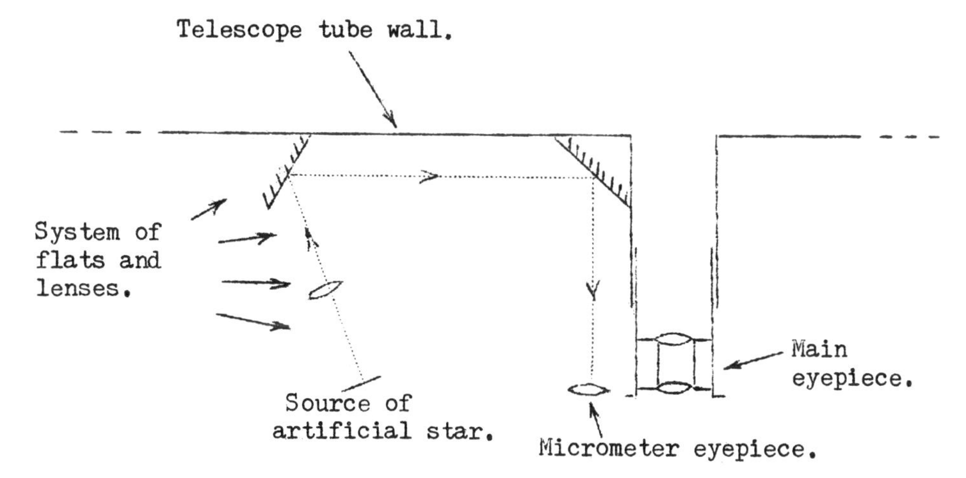

Fig.11. Method of projecting artificial stars into left eyepiece.

Fig. 11 shows how the artificial double star images are projected into the lefthand eyepiece of a pair of "binoculars". There are three different ways in which the artificial double star images may be produced.

Micrometers for Double Star Measurement.

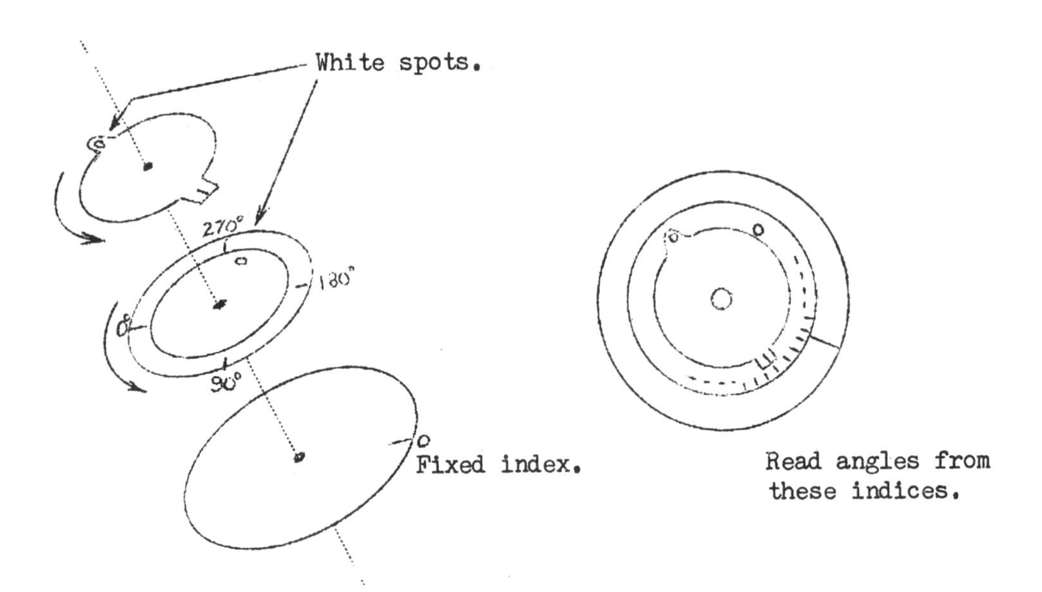

Fig 12. The separate components of the micrometer compass. The upper two components rotate about the axis.

Fig 13. View of components in Fig 12 assembled. The white spots are projected into the micrometer eyepiece.

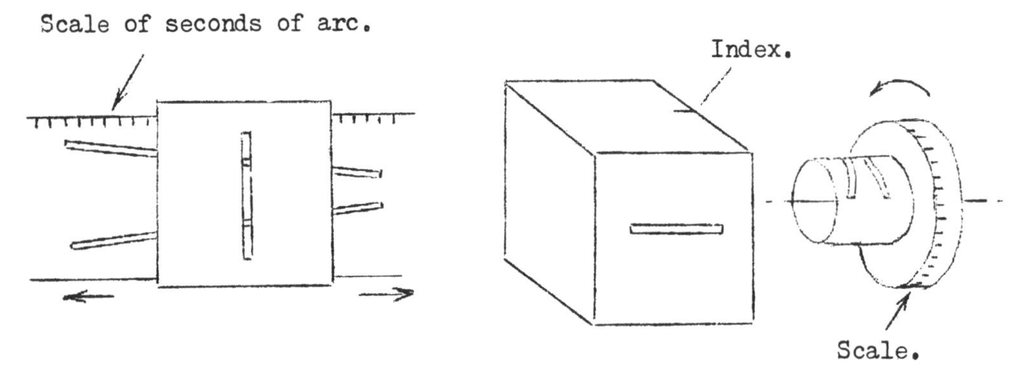

Fig 14. The image formed by the slits with rear illumination forms the artificial star.

Fig 15. A similar method to Fig 14 with a rotating scale. A small lamp is housed inside.

Micrometers for Double Star Measurement.

(i) A pair of compasses is used with a white spot representing a star on each branch. This is seen against a dark background, preferably black velvet, and the position of each branch is read off from a $0^\circ - 360^\circ$ protractor. If the angle between the branches is β and the distance between the two artificial stars is d, then the position angle is given by $(\beta/2) + 90^\circ$ where $\beta = (\theta_2 - \theta_1)$ and θ_2, θ_1 are the two angles "pointed at" by the branches. The separation d is proportional to sin $(\beta/2)$ and can be calibrated by measuring several standard pairs and plotting d against sin $(\beta/2)$. Hence when a new pair is measured a knowledge of $(\theta_2 - \theta_1)$ will yield d directly.

It is useful to have several sets of compasses having a pair of bright stars, a pair of faint stars and an unequal pair to prevent systematic errors. By using an optical system as shown in Fig.11 the scale can be increased so that 1" of arc could be represented by a distance of several mm between the two artificial stars.

(ii) A second method is to cut two thin lines which are slowly converging in a length of paper. If these are illuminated from behind and viewed by means of a perpendicular thin line, also cut from paper (Fig. 14) then when the paper containing the two converging lines is moved, two artificial stars are seen which vary in their separation.

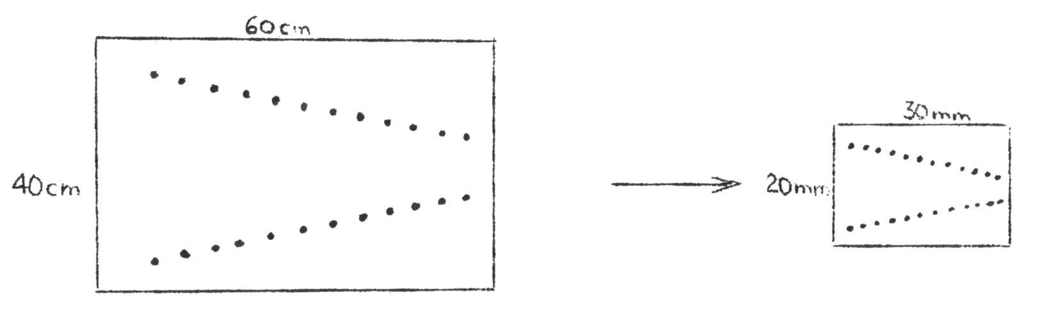

Fig 16. Artificial pairs of known separation, drawn as black stars on white background and photographed to produce clear images on black background.

(iii) If a number of "double stars" of varying separation are drawn on paper (Fig.16) and photographed, this can be reduced and projected into the lefthand eyepiece in a similar manner to the first two methods.

The errors in these methods are close to those inherent in the filar micrometer. The advantages are that the whole aperture of the telescope is available and that no field illumination is required making fainter pairs accessible to the observer.

Micrometers for Double Star Measurement.

(d) THE DIFFRACTION MICROMETER.

The appearance of a bright star in a small telescope is that of a small, brilliant point of light. If a coarse diffraction grating is placed over the object glass or mirror, then a succession of fainter images will be seen extending linearly from the star image in a direction perpendicular to that of the grating slits and becoming fainter as they do so. The number of satellite images visible depends on the brightness of the star and the number and width of the slits in the grating.

If two stars are seen close together in the eyepiece then each star will be accompanied by its own system of images. By rotating the grating the images can be arranged into certain patterns which allow the observer to measure the distance between the two stars and to determine their position angle. Fig.17 shows the appearance of a double star with the grating slits (a) horizontal and (b) vertical.

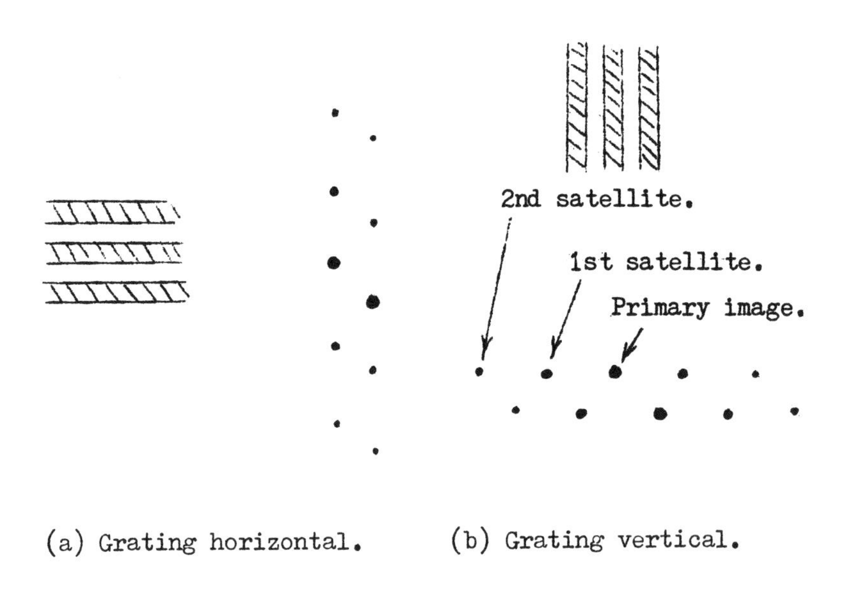

Fig. 17. Appearance of a Double Star through a grating.

Each satellite image is equidistant from the other by a distance of z seconds of arc, where:

$$z = \frac{206265\lambda}{p} \quad \text{--- (1)}$$

p being the total width of the grating bar and slit and λ is the mean wavelength of the incident light. (In a pair of stars where the components are of different spectral classes, the value of z is in fact different for each star, but the difference cannot be detected visually). It is sufficiently accurate, within 1% to determine ones own value of mean wavelength.

Micrometers for Double Star Measurement.

This can be done in two ways:

(i) By measuring a well known pair whose separation and position angle are well defined. Such pairs are eta Cassiopeiae (STF60), gamma Virginis (STF1670) and xi Bootis (STF1888).

Table 4 gives their separations and position angles for the years 1979-1981.

Table 4.

Separation (")	STF 60	STF 1670	STF 1888
1979.0	11.94	3.97	7.23
1980.0	11.99	3.90	7.22
1981.0	12.03	3.82	7.21
Pos. Angle (°)	**STF 60**	**STF 1670**	**STF 1888**
1979.0	306.5	297.6	333.8
1980.0	307.0	296.9	333.0
1981.0	307.6	296.1	332.3

To obtain position angles and separations for dates in between the observer could use linear interpolation without incurring any appreciable error. The figures are taken from the Third Catalogue of Double Star Ephemerides by Muller and Meyer, published by the Observatory of Paris. With an accurately known separation and knowing z, the observer can find λ although many measures should be made before adopting a mean figure to use in later observations.

Table 5. Values of z for various values of p (λ = 5560Å).

z"	p (cm)	z"	p (cm)
11·47	1·0	1·91	6·0
7·65	1·5	1·76	6·5
5·73	2·0	1·64	7·0
4·59	2·5	1·53	7·5
3·82	3·0	1·43	8·0
3·28	3·5	1·35	8·5
2·87	4·0	1·27	9·0
2·55	4·5	1·21	9·5
2·29	5·0	1·15	10·0
2·09	5·5		

(ii) For a second method of determining ones personal mean wavelength λ, it is necessary to observe a star of high declination, preferably Polaris. The proceedure is to use a grating whose value of z is fairly large, (M. Duruy uses 7" of arc) and to orient the grating so that the images are aligned E - W in the stationary telescope. An eyepiece with a single wire is rotated so that the wire is N - S and the time (t) between images A' and A" crossing the wire noted, (Fig.18).

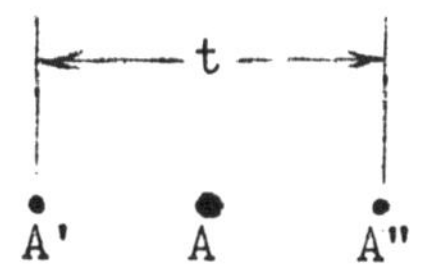

$t = 2z/15\cos D$

where t is in seconds of time, z is in seconds of arc and D is the declination of the star.

Fig. 18 Eyepiece wire method for obtaining λ.

In the case of Polaris where $D = 89^{\circ}$ the time required for A' - A" to cross the wire is about 53·5 seconds. Using a stop watch, this time can be determined to within about 0·2 second and this corresponds to a mean error of the order of 0·3%. Thus from equation (1), knowing z and p, λ can be found.

Ideally, there should only be two satellite images visible for each component of the double star, making six in all. When bright stars are examined, there are many satellites and the brighter ones are short spectra, and hence rectangular in shape. M. Duruy, using a 24-inch reflector found that in that instrument, satellite images of stars fainter than magnitude 3 do not show up as spectra. For a 6-inch reflector this corresponds to a magnitude of 0, thus the vast majority of pairs will show rounded images, since the images are too small and faint for the eye to distinguish between a square and a circle.

The number of satellite images can be controlled by means of the grating slit and bar width. M. Duruy has studied the variation of the faintness of successive satellites, which depends on the ratio a/p, as shown in Fig.19.

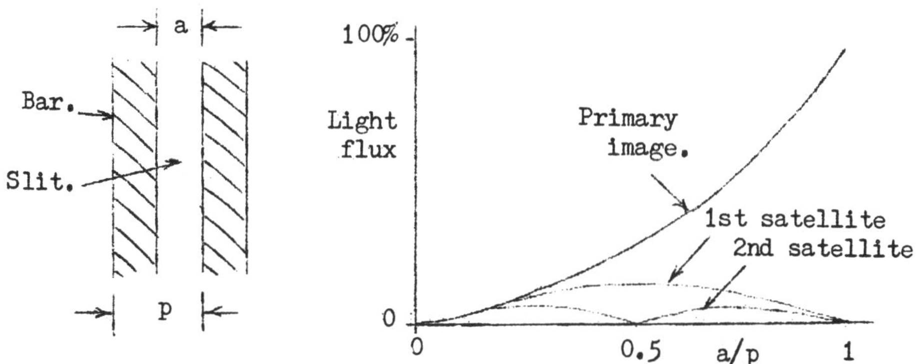

(a) Grating nomenclature. (b) Variation of transmitted light plotted as a function of a/p.

Fig 19. Satellite image faintness and ratio a/p.

Micrometers for Double Star Measurement.

When $a/p = 1$, the bars disappear and there is no diffraction. When $a/p = 0$, no light passes through and it was found that for $a/p = 0{\cdot}5$ (i.e. slit width equal to bar width) the 2nd satellite disappears whilst the 3rd and higher are usually too faint to be seen. Thus the optimum value of a/p is 0·5 and this was adopted by M. Duruy for his micrometer.

The grating and position circles should be so arranged that when the pointer (parallel to the slits) shows a P.A. of 90° or 270° the images are horizontally aligned and hence parallel to the direction of drift of an equatorial star with the telescope stationary. The pairs STF 1657 (24 Com), P.A. = 270° and STF 180 (gamma Arietis), P.A. = 0° can be used to check this setting.

To measure the separation of a pair there are two main configurations which can be used.

(i) The method of right angled triangles.

The grating is turned until the angle ABA" as in Fig.20(a) is adjudged to be a right angle. The reading on the circle, α_1, corresponding to this position is noted. Then the grating is moved in position to give the configuration as in Fig.20(b), and the angle α_2 noted.

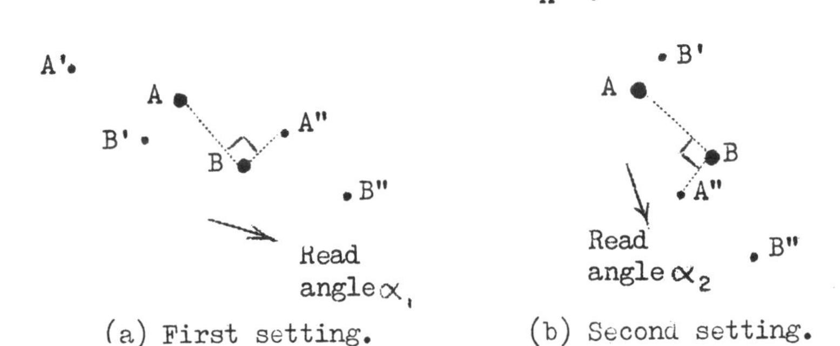

(a) First setting. (b) Second setting.

Fig.20. Configurations of images for pair separation, method of right angled triangles. ABA" = 90° at each setting.

If $\alpha = \alpha_1 + \alpha_2$ then ρ which is the distance AB is given by:

$$\rho = z(\cos\alpha/2)$$

where z is known and hence ρ, the distance of the pair can be found.

Also, $\theta = (\alpha_1 + \alpha_2)/2$ where θ is the P.A. of the pair.

(ii) The method of isoceles triangles.

In this method the grating is successively rotated to give the configurations as in Fig. 21(a) and (b), the criterion of setting being that AA' = BA' and that AB" = BB".

α_1 and α_2 are read as in the first method and ρ and θ can be obtained from:

$$\rho = 2z(\cos\alpha/2) \qquad \text{and} \qquad \theta = (\alpha_1 + \alpha_2)/2$$

where $\alpha = \alpha_1 + \alpha_2$ as in the method of right angled triangles.

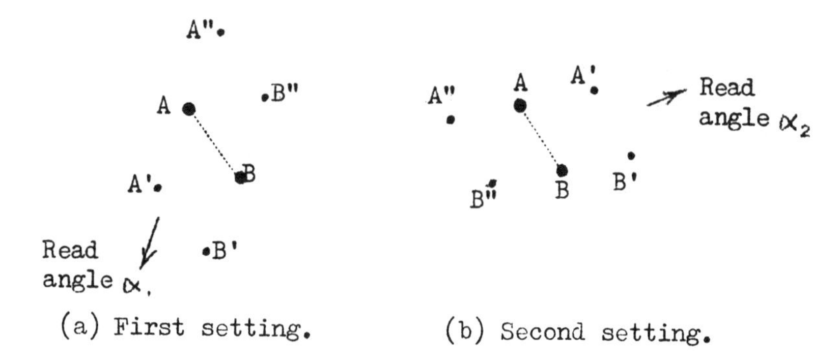

Fig.21. Configuration of images for pair separation, method of isosceles triangles. AA' = A'B = AB at each setting.

Hence with the same grating, two different methods can be used to find the position angle and separation of a pair. It is better to make a large number of measures of each pair, preferably using both methods. By constructing several gratings of different values of d it is possible to cover quite a large range of separation.

Another method of using the grating requires a single wire in the eyepiece and a position angle circle to read the orientation of the wire. This is shown in Fig.22.

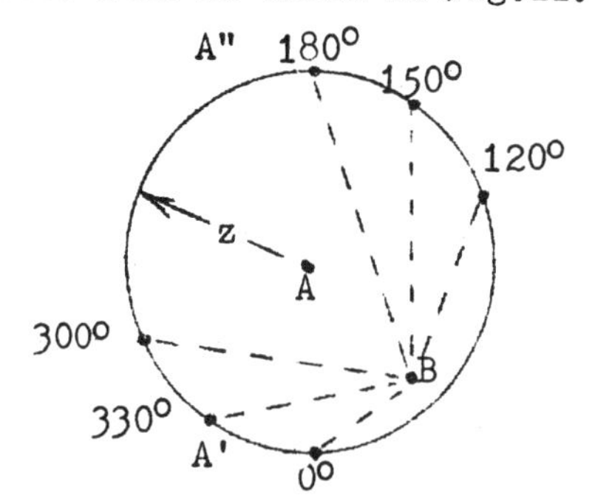

The radius of the circle is z. The three points A" are at known angles which are set up on the grating circle. By measuring three angles A'B and three further angles A"B, construction of the diagram shown to a fixed scale will give the value of ρ and θ.

Fig.22. Configuration of images, eyepiece wire method. ($\rho < z$).

In this method, the value of z is greater than the separation of the pair. The grating is oriented so that the secondary images lie in P.A. = 0°, i.e. N - S. The wire is turned so that it measures the angle between A' and B' and this reading is noted together with the orientation of the grating. Then angle A"B is measured and noted and the grating then turned so that the images lie in a different known angle, say 330°. Again the angles A'B and A"B are taken and noted, the drawing as in Fig.22 being reproduced on paper at the desk later. By drawing it to scale, the position angle and separation can be obtained geometrically.

A similar method exists for the case when z is less than the separation of the pair. It has been found by M. Duruy that in this case the position angles are better defined than the separations: when z is greater than the separation the opposite is true.

The main disadvantage of the diffraction micrometer is the loss of light suffered due to the grating bar area. When a/p = 0·5, only 50% of the incident light from the pair will reach the mirror or object glass. M. Duruy has calculated that 25% of the incident light goes into the primary image making it 1·5 magnitudes fainter than without the grating. Of the remaining 25%, 10% goes into each of the first order images which are thus 2·5 magnitudes fainter, whilst the remaining 5% is spread amongst images of higher order.

Two instruments using the same principle as the Duruy micrometer appeared previously. In 1895, K. Schwarzchild utilised a grating over the object glass which was hinged down the middle, both halves being tilted with respect to the object glass by means of a rack and pinion arrangement. This allowed a variable value of z but involved the measurement of another angle. Lawrence Richardson, in various BAA Journals, described a rectangular coarse grating which was tilted on a pivot in front of the object glass. Both of these devices are more elaborate than the Duruy micrometer.

I must take this opportunity of expressing my thanks to M. Duruy who has made freely available to me all his notes and comments collected over 10 years of micrometrical work with this device.

(e) INTERFEROMETERS.

The application of interferometers to the measurement of close double stars was primarily due to the work of Albert Michelson. In 1890 he suggested that interference methods could be used to measure the separation and position angle of close pairs. However, it was not until 1920 that Pease and Anderson, using a simple two-slit device, measured Capella with the Mount Wilson 100-inch reflector (Fig.23).

If a star is observed using the two-slit device, its image will appear as an enlarged and elongated diffraction disk, crossed by a number of bright and dark fringes. These fringes are a result of the interference of the light from the two slits and the width of the fringes depends on the wavelength of the incident light.

Micrometers for Double Star Measurement.

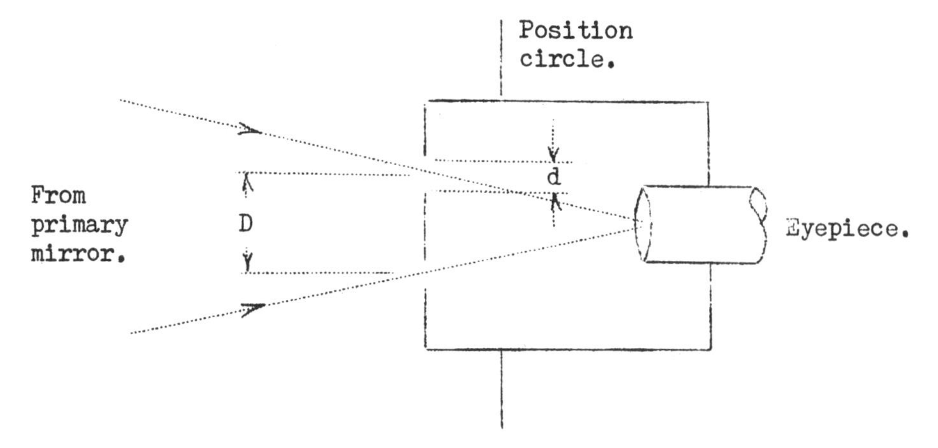

Fig.23. The Two-slit Interferometer.

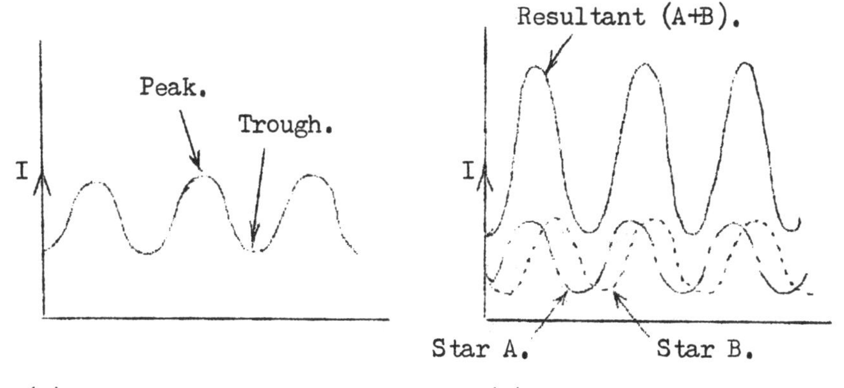

(a) For a single star. (b) For a double star.

Fig.24. Fringes produced by a Two-slit Interferometer.

If the star is a very close double, each component will produce its own set of fringes in the eyepiece (Fig.24(b)), the separation of the two sets depending on the distance between the slits D. If the value of D is changed, the fringe patterns will move relative to each other and if the peaks in each system co-incide with each other, the result will be a brighter set of fringes. If, however, the peaks of one system and the troughs of the other are co-incident, there will be cancellation of the fringe system, the fringes disappearing completely if the components are equally bright. For the smallest value of D at which the fringe visibility is a minimum:

$$\alpha = \lambda/2D$$

where α is the angular separation of the two stars and λ is the effective wavelength of the incident light. This value of α is 2·44 times smaller than that for a normal telescope, but due to practical difficulties the effective gain is about 2·0, i.e., the lower limit of resolution for the 100-inch reflector is 0"·025 with the interferometer and about 0"·05 without.

When the slits are rotated about the optical axis in their own plane, the fringes disappear at certain angles. This is because the bright fringes of one component are in exact register with the dark fringes of the other but they are laterally displaced and instead of the normal minimum being formed, a zig-zag pattern is seen. This pattern occurs four times in one revolution of the slits, forming two sets, 180° apart. Knowledge of these angles allows the position angle of the pair to be found, although one extra reading will be required to find in which quadrant the comes lies.

A modern application of this method has been developed by Dr. W. S. Finsen, of the Republic Observatory at Johannesburg, R.S.A., who constructed an eyepiece interferometer in 1933 and made an improved version in 1948. Dr. Finsen has made many thousands of measures with this device and is thoroughly acquainted with the many fringe patterns that are possible. He has also made extensive investigations into the effect of atmospheric dispersion on the fringes. This has the effect of shifting the centre of the fringe pattern away from the centre of the diffraction pattern at certain position angles. Other effects which complicate the fringe pattern include the fact that starlight is a spread of wavelengths, and also that the pair might not be equally bright. It has been found that measures cannot be made effectively for pairs differing by more than about one magnitude because the fringe contrast becomes very indistinct. The measures which Dr. Finsen has made include pairs whose separation is less than 0"·1 and they are valuable because they are more reliable than ordinary micrometrical measures at such small separations.

A new application of Michelson's work has proved to be a very powerful tool in the measurement of extremely close pairs, down to perhaps 0"·01. This method makes use of the fact that even in stellar images affected by the turbulence in the Earth's atmosphere, there is useful information limited only by the diffraction of the telescope optics. A highly magnified short exposure of a bright star in a narrow band of wavelength will show a large number of small grains in the image. This effect is known as the speckle pattern, and is due to random interference between the starlight in the Earth's atmosphere. The minimum size of the speckle grains is equal to the size of the Airy disk for the telescope. If the star is a close double, there are two identical speckle patterns superimposed, and shifted relative to each other by an amount smaller than the Airy disk. If the negative of a speckle photograph is illuminated with a laser, interference fringes will be seen, the separation of which is proportional to the separation of the double star. By combining a number of short exposures a better fringe pattern is obtained. The orientation of the fringes yields the position angle, whilst the contrast is a function of the magnitude difference of the pair. From photographs taken with the 200-inch by Gezari, Labeyrie and Stachnik, it was found that beta Cephei (magnitude 3·3) has a very close companion of magnitude 9·0 at a distance of 0"·25, a star which is invisible by visual means. Other photographs in Sky and Telescope show resolution of the binary Capella which has a maximum separation of 0"·055.

In conclusion, the instruments described in this chapter may be divided into two groups of different difficulty of construction.

The simplest devices are the diffraction micrometer and the binocular micrometer that utilises a pair of compasses. The former can be made of wood or cardboard, but aluminium is to be preferred. It is important that the grating is constructed with accurately parallel and equidistant bars. Since all measurements are made using the position circle it is important that this is accurate and easily read. It should not be too small, and a minimum of 12 inches diameter is recommended. This will have degree markings approximately 0·1 inches apart. This also applies to the binocular micrometer, and it should also be noted that with this instrument the scale constant is dependant upon the magnification and should be determined separately for each eyepiece.

The double image, comparison image and filar micrometers, also the eyepiece interferometer all have to be accurately engineered. The last is of limited application since it will only operate on pairs where Δm is less than one magnitude and its range of separation measurement does not approach that of the other instruments in this group. Of the remainder, both the Muller version of the double image micrometer and the comparison image micrometer require birefringent prisms which are expensive. Once installed, however, these devices are more stable than the filar micrometer where wires may break and have to be replaced.

The Author has no experience of using the eyepiece interferometer but has found that the comparison image micrometer is more comfortable to use than the others in this group.

For the amateur with little engineering capacity available the diffraction micrometer is strongly recommended as a first instrument to construct for the measurement of double stars.

6. PHOTOGRAPHY OF DOUBLE STARS.

(a) THE PHOTOGRAPHIC PROCESS.

Before discussing the photography of double stars it would be as well to re-capitulate upon the photographic process generally. There are many methods of writing-by-light, and in this respect any change of colour or density by the action of light might be termed photography.

The particular process that we are interested in is that which utilises a suspension of silver halides in gelatin spread onto a transparent base, either glass,(in which case the product is termed a plate), or a flexible material, now usually cellulose acetate, (in which case the product is called a film).

This process has developed from the original work of Fox Talbot, who, more than 100 years ago first successfully produced a 'negative' picture with light and shade reversed, and from which a number of subsequent copies could be taken. His process was complex and somewhat hit-and-miss, and it is a good thing that we have available the emulsions, (as the suspensions of silver halides are termed), that science has developed for us.

The modern emulsion consists of a very carefully prepared suspension of silver halides, together with sensitising agents etc., produced to a quite remarkably consistent quality and available to meet practically every photographic need.

Early emulsions were only sensitive to the blue end of the spectrum, hence the use of the classic red darkroom lamp for their processing. Then methods were found to extend the range of sensitivity into the yellow and green, giving 'Orthochromatic' emulsions , and later, with the aid of further sensitising dyes, panchromatic (or 'Pan') materials were produced. These respond to the entire visible spectrum, although, except in the case of some very special films, they are still more sensitive to blue light. In addition, virtually all films respond to some degree to the ultra-violet.

Modern processes also allow us to control the 'speed' of emulsions, that is, whether it will take a lot or a little light to have a given effect in a given time. Many systems of quoting these speeds have been used, those in common use now being the DIN system, which is logarithmic and the ASA system which is arithmetic. It is not proposed to discuss these in detail; they are covered adequately in any book of basic photography. Suffice it to say that the slowest films have a speed of about 6 ASA whilst the fastest that can be attained with special processing is 10,000 ASA or even higher, depending on the method used in measuring the speed. With the comment that faster films generally have coarser grain structure, and cannot therefore resolve such fine detail, the discussion of the nature of emulsions will be left to consider what happens when a photographic emulsion is exposed to light.

Photography of Double Stars.

As we are considering stellar photography (and stellar images are essentially point images) let us consider a photographic emulsion that had previously been un-exposed to light, and upon which a point of light of very small dimension is allowed to fall. The silver halides are in the form of very small crystals, and for reasons still not completely understood, the action of light is to produce in them what is called a 'latent image'. This means that if, subsequent to exposure the emulsion is acted on by a suitable reducing agent, or DEVELOPER, the affected crystals of silver halide will be reduced to metallic silver, which, in the finely divided form it takes, appears black.

If we now use a further chemical, called a FIXER, which is capable of dissolving any undeveloped silver halide, but which does not affect metallic silver, then we may, provided certain precautions are taken, let our emulsion see the light of day and we shall have a permanent record of our point of light as a dark spot on an otherwise clear background. If the point of light had been the image of a star in a telescope, then we now would have a permanent record of that star.

So far so good. The foregoing has necessarily been simplified, and in order to undertake stellar photography, (including double stars), there are a number of other factors to be taken into consideration. Those that have a bearing on the matter are as follows:-

(i) Film Speed.

The speed of the film, or more correctly, the speed of the film/developer combination, since these are inter-related. Obviously the faster the film, the fainter the star it will record in a given time.

(ii) Image Size.

Some light will be reflected from the crystals of silver halide, and will illuminate adjacent crystals, which will become developable. The image will then be larger, (many times, perhaps scores of times larger for bright stars), than the original image, even assuming that this was perfect in the first place. Imperfect seeing, either wandering or boiling, will enlarge the image further.

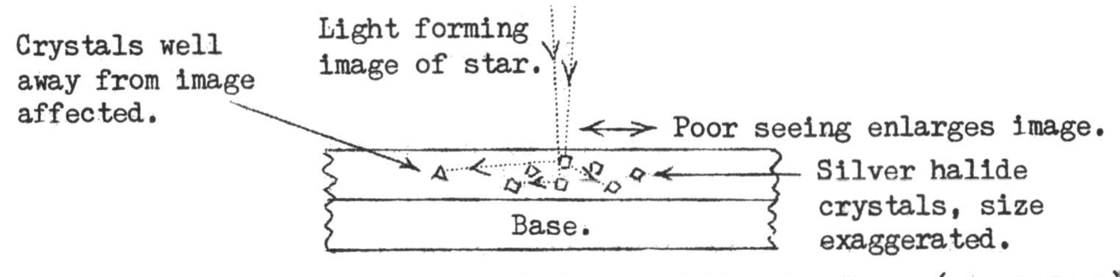

Fig 25. Formation of photographic image of stellar disc, (simplified)

Photography of Double Stars.

(iii) Failure of the Reciprocal Law, or 'Reciprocity Failure'

Manufacturers usually quote the speed of their emulsions for exposures in the region of 0·1 second. In theory, if a given amount of light gives a certain degree of developability in a given time, then half that amount of light should have the same effect in twice the time, and so on. This relationship fails for very low levels of illumination, when long exposures must be used, and for extremely short (micro-second) exposures in very intense light. In each case the emulsion seems to be slower than expected.

The micro-second exposures do not concern us, and for the purpose of such double star work as will be discussed, the effect of low level illumination and the resultant exposure times will not have sufficient effect to be appreciable. Fast films generally suffer more from Reciprocity Failure than slow films. As examples, materials with speeds in the order of 400 ASA or higher will start getting 'tired' by the time an exposure of 30 minutes has been reached, whilst slower films, around 25 to 50 ASA will continue at their rated speed for a couple of hours or more.

(iv) False Reciprocity Failure.

This occurs with very bright images, and should not be confused with true reciprocity failure, which occurs with all images and is a function of exposure time.

Briefly, what happens is as follows. A bright image will rapidly affect all the crystals of silver halide that it falls upon, and although these affected crystals will continue to reflect light to adjacent crystals so causing the size of the image to expand, there can be no further action on the central crystals. Any light absorbed by them is lost so far as producing an image is concerned. This is of especial importance when the image diameters are used to assess the magnitudes of the stars forming them. It would be safe to say that any star more than about two magnitudes above threshold (i.e. that star brightness which has _just_ produced a fully dark circular image) will start to exhibit false reciprocity failure.

(v) Background Darkening.

No sky is completely dark, and the rate at which the general background will affect the film will depend upon the focal ratio of the system. For this reason refractors are better than reflectors, although by adding a 2x Barlow of suitable quality, the situation can be remedied. Subject to certain minor considerations which do not affect the matter sufficiently to merit inclusion, the faintest magnitude reached will be a function of linear aperture, as in visual work, whilst the degree of darkening will be a function of focal ratio.

Photography of Double Stars.

Thus a 15-cm reflector of F/8 will record to a given magnitude on a given film in a given time. A 15-cm refractor of F/16 will reach the same magnitude in the same time under the same conditions, but can continue four times as long before the background has darkened to the same degree. The linear scale will also be twice as large, but this will be discussed in greater detail later in this chapter.

The above are the chief factors that affect the behaviour of the photographic emulsion in the recording of stellar images, it being assumed that these are being taken at the focal plane of the optical system. So far as double stars are concerned, photography cannot be considered the most accurate method of recording and measuring - the eye and the micrometer must take first place here.

Two déficiencies that occur in the photographic process are as follows. Assuming perfect seeing, perfect guiding and perfect optics, the developed images of a double star ought to be two circular patches of silver of a size determined by the optics, film speed, exposure and magnitude, and of a spacing determined by the separation of the stars and the geometry of the optics. Unfortunately this is not so. Firstly, there are random movements of the order of 10 - 20 microns in the gelatin of the emulsion during processing that cannot be predicted, and secondly there is an effect in which the by-products of the development of the image have a restraining effect in their immediate vicinity, and this can distort the shape and even the position of closely adjacent images, possibly partly inhibiting their full development. Each of these factors detracts from the usefulness of a photograph of a double star for the purpose of accurate measurement, although with wider pairs and a long focal length, results are not far short of those obtainable with a visual micrometer.

(b) CHOICE OF MATERIALS.

It might seem that a very fast material would be the best to use, as exposure would then be short and the accuracy of guiding need not be of such a high order. In practice this is not the case. If a stellar image on a piece of film is examined under a microscope at a magnification of about x100, it will be seen that the edge of the image is indeterminate, gradually shading into the clear background. The faster the film, the coarser the grain and the less determinate the form of the image. On the other hand, very slow films will give images with beautifully crisp edges but are rarely circular due to difficulties in guiding for the required long exposures.

A good compromise is to use a medium speed film, such as Ilford FP4 which has a medium fine grain, a good panchromatic response, and if used with Hyfin developer may be taken as having a speed of 200 ASA. This is not spectacularly fast, but Hyfin developer gives crisp sharp images. There is nothing to be gained in terms of finer grain by using anything slower. If a higher speed is required, then Ilford HP4 can be used, being processed in Acutol-S,(for optimum speed/grain characteristics), or Acuspeed, (for maximum effective speed). Use these developers in accordance with the manufacturers directions. Their use at lower concentrations for longer periods of development will not increase the effective speed, although for bright stars there will be less background darkening. In this case, however, such a fast film is not really required. In addition, the effects of reciprocity failure seem to occur earlier when the dilute developer/extended processing time procedure is followed.

It might be as well to mention that it is useful to be able to load one's own cassettes with bulk film in short lengths, thus allowing the film and its subsequent processing to be matched to the subject. Beware of some of the very cheap film that is advertised, as this is sometimes cine film, and is very subject to chemical fog, which is the random reduction of silver halide crystals to silver independant of the effect of light on the emulsion.

Always buy fresh material, it is not expensive in lengths of 5 and 17 metres, and also in 100 feet lengths from some supply houses. This will cut your film costs to less than one half of the cost of ready loaded cassettes and allow you to use just that amount of film that is needed. If you have no cassettes, your dealer will be able to supply for a few pence each, but make sure that they are in good condition and that the velvet light traps are clean and free from dust. A clean discarded toothbrush is useful for this. In any case, discard a cassette when it has been used about 6 times, it has repaid its cost by then. A bulk film loader such as that manufactured by Burke and James, and sold under the name of 'Watson' is a very useful accessory for loading film if you have much work to do, but is by no means a necessity.

Similarly with chemicals. Buy these fresh and make them up according to the manufacturers instructions. Above all else, do not store film and chemicals together, something or other unpleasant is bound to happen. With a bit of luck, a suitable corner of the home can be found that is cool and dry for storage. Remember that photographic chemicals are not for consumption, so no developer or fixer in old gin bottles in the sideboard, and keep all film and chemicals well away from children. Whilst film is not poisonous, the sight of a three-year old playing yo-yo with 5 metres of previously unexposed FP4 can promote the most unpredictable behaviour in the quietest astronomer.

(c) APPARATUS AND EXPOSURE.

The range of apparatus that might be used to photograph double stars is so wide that it would be impracticable to attenpt to cover every possibility. For those who have not attempted any stellar photography before, the following table may prove useful. All of the data in it is taken from exposures made by the writer, not specifically for double star work, but it will give a fairly good idea of what may be expected from various instruments. In each case the star taken as the example is approximately 2 magnitudes above the absolute threshold of the exposure. The film used was FP4, processed in Hyfin.

Table 6. Magnitude and Image Diameters with various Apertures.

Objective and F.L.		Lin. Aperture.	Exposure.	Mag.	Dia, (" of arc)
Telephoto lens.	135mm.	34mm.	2 minutes.	7	45
-ditto-	400mm.	63mm.	2 minutes.	9	21
Speculum.	1400mm.	216mm.	1 minute.	10	7
Object Glass.	3200mm.	204mm.	2 minutes.	11	3

The above figures are not completely comparable, as the observing conditions were not necessarily identical on each occasion, but they will suffice to indicate the results to be expected. If anything, the magnitude figures are conservative.

The limiting magnitude that might be expected is dependant on many factors, not least of which is the sky brightness. Obviously the best combination is a large aperture, long focal length and accurate guiding. This will give a large scale, minimise sky brightness and give access to fainter stars. Unfortunately, to meet these requirements the accuracy of guiding needed is high, and needs an equatorial head that is in accurate adjustment and a drive with very smooth operation and control. Except near the zenith, changes in refraction may well affect the result; as the last entry in the above table illustrates, with an image only 3" across, an error of only 1" will produce a decidedly elliptical image. This effect can be minimised by making exposures as near the meridian as possible.

As a general guide, with a head in good adjustment, and good driving apparatus, with smooth differentials, or frequency control in the case of variable frequency devices, it ought to be possible to give an exposur in which the product of focal length in mm and the exposure in minutes is anything up to about 18,000 with little guiding correction, if any. According to the accuracy of setting of the head, nearness to the meridian and smoothness of control of drive, this figure can, of course, be considerably exceeded, although it must be remembered that we are only considering guiding problems. Sky brightness may well prove to be the limiting factor.

So far as the type of exposure that may be given, the simplest is to expose in the focal plane, which will record the double as two dark circular images on the film. If one wishes to make some measurements, then it is necessary first to determine the precise scale at the focal plane, and for this purpose the Pleiades are probably the best object in the sky. They are bright, so that exposures can be short, and guiding problems are at a minimum, and their positions are known accurately. Failing this, photograph several well-known fixed doubles, and obtain the scale from the resultant exposures. If you use a Barlow lens, be sure that you can replace it in precisely the same place each time. Measurement of the exposures, either with a compound microscope with eyepiece micrometer, (a simple graticule is sufficient), or by projection with a lens of known good linearity (probably an enlarging lens of the highest grade) will give the separation, provided that the centres of the images are estimated accurately. Each measurement should be made several times and an average taken.

If the exact effective focal length of the optical system in use is known, then the image scale may be determined from the formula.

$$s = 206265/f$$

where f = focal length in mm and s = scale at the focal plane in seconds of arc per mm.

No reference is made in this section to the means of placing the film in the focal plane. Much has been written elsewhere, and the reader is referred to the bibliography at the end of this book. However, the single lens reflex system, or some other arrangement that allows the observer to see the image about to be photographed has so much to recommend it that it may be considered almost an essential. This does not mean that good work cannot be done without such a system, but the ease of working means that much more work can be done. Suitable single lens reflex bodies may be obtained secondhand very reasonably, and the addition of a suitable lens at a later date will provide the observer with a general purpose camera.

The type of exposure that has been discussed so far is merely that of the images of the two components of the double, together with the images of any other stars in the field. To determine position angle two courses are open. The first is the more accurate if one of the stars is bright enough to leave a trail with the telescope drive stopped. In this case, with the driven exposure completed, (the star having been placed towards the following side of the field), the exposure is interrupted for, say, 5 seconds, with a dark card over the aperture with the drive stopped. The card is then removed and the star allowed to trail out of the field.

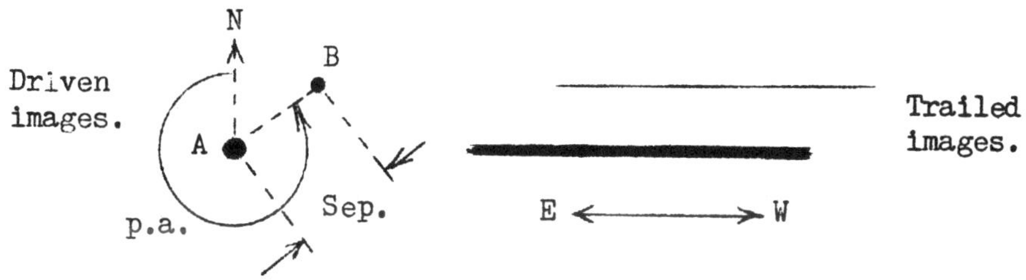

Fig 26. Measurement of Position Angle. (Trail Method).

Photography of Double Stars.

Unless the exposure is very close to the Pole, or the field of view very large, the result will be to all intents a straight line which will define the E - W direction. Make sure that the trail will not obscure either of the images, as may occur where the position angle is near 90 or 270 degrees and also try to assess the effect of the trails that will be left by any other stars in the field.

To get a brighter image, do not merely slow down the telescope drive. Unless this is perfectly in an E - W direction, the result will be to compound the movement of the telescope with the diurnal motion of the star and give a false E - W direction.

If neither star is sufficiently bright for the preceding method, after having made the first exposure, and with the exposure interrupted with the drive stopped, restart the drive and make a second exposure as far from the first as the field will allow. A line joining the same star in the double exposure will define the E - W direction. This method is very dependant on having a telescope mounting that can be stopped and started precisely, and with absolutely no movement in N - S direction during stopping and starting, since this will give a completely wrong result.

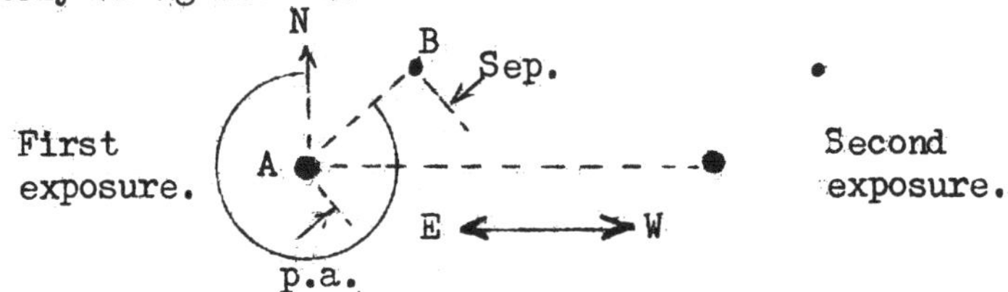

Fig 27. Measurement of Position Angle. (Double Image Method).

With reference to guiding, it ought not to be necessary to correct in declination for exposures of the order of which we are speaking. Unless such corrections are made continuously the result will be painfully apparent. The answer lies in getting the equatorial head into really accurate alignment, and eliminating any flexing or movement between the camera and the guide instrument. It is preferable to make this guide instrument adjustable, so that it can be offset to guide on an adjacent bright star. This can be thrown out of focus and the resulting circular image kept carefully quartered on the crosswires of the guiding eyepiece, which in this manner of use require no illumination. It may be noted that with a guide instrument of 11-cm aperture a magnification of x40 is suitable when the camera in use is fitted with a telephoto lens of 400mm focal length, and of x225 when making exposures at the prime focus of a 216mm speculum of basic focal length 1,400mm used in conjunction with a x2 Barlow lens. Stars down to about magnitude 7 are sufficiently bright to be thrown sufficiently out of focus to be suitable for guiding.

One remaining field of exposure may be mentioned, although the writer has no personal experience of it. This is the combination of the diffraction micrometer with prime focus photography. Since the measurements to be made are angles, the method might well commend itself to an observer who wishes to make as accurate measures as possible with photography, although the factors that affect the image positions, shifting them from their theoretically correct positions, would have to be minimised by using a fairly long focal length.

There is one further possible advantage that the use of a simple objective grating could have, and that is the determination of the position of the image of the brighter component of a very unequal pair. It would be easier to find the midpoint between the two first-order spectra than to estimate the centre of a grossly over-exposed image. This method has been used in studying the movements of the components of Sirius, allowing the position of Sirius A to be decided more precisely.

(d) CONCLUSIONS.

All astronomical photography is more or less experimental, and this applies more to the photography of double stars than to any other field. It is necessary to make test exposures, to determine the magnitude limits attainable with your equipment, the scale of the image and the effect of sky darkening. Remember that a Barlow lens will help in the last case with a reflector of short focal ratio. If your funds will run to it, and assuming that you are using a single lens reflex camera body, the use of a 'tele-extender' of x2 has the double advantage that it will be a highly corrected lens, (although being usually of 4 element construction will introduce a certain light loss), and that if used with the correct adaptors to suit the eyepiece focussing mount and the camera body, will always occupy the same position thereby providing a constant scale.

In general, filters are not of great use in this type of work, but in one case the use of the correct filter can be invaluable. Increasing areas are being illuminated by sodium lighting and this constitutes a major factor in background darkening on exposures. However, except in the case of the high pressure type of lamp, (which has light of a pinkish tinge), almost all of the light is contained in a very narrow band of the spectrum in the yellow, (the 'D' lines). The use of neodymium filters, also called didymium filters, which are opaque to a narrow band including these lines can eliminate this, but such filters are comparatively expensive and are fairly opaque to the remaining light, necessitating long exposures. A better alternative is to use a blue filter, such as the Kodak Wratten 44 or 44A. The latter is preferable, since it has a better transmission up to the cut-off point, which is just short of the 'D' lines, but is usually a special order item. Both of these filters are available in gelatin, and are inexpensive. They must be handled carefully and not allowed to become damp or dewed up, but if they do suffer damage they are cheap to replace. Either the 44 or 44A will reject more than 99% of sodium lighting, whilst exposures require increasing by about 1.7 for the 44A and around 2.2 for the somewhat denser 44. It will therefore be seen that there is an effective gain of not less than 4 magnitudes over the background if this is only due to sodium lighting.

The use of a blue filter alters the spectral response of the emulsion, of course, and it is the experience of the writer that either of the filters mentioned, used in conjunction with either FP4 or HP4 film will result in images being recorded in a similar manner to that which would obtain with a non-colour sensitive emulsion as used in the early days of photography, and the classic colour index applies sufficiently closely.

So far separation attainable has not been mentioned. With due precaution and a sufficient focal length it should be possible to separate photographically two equal stars in the 'two magnitude above threshold' range which are of the order of 2 or 3 times the Dawes's limit for the aperture. The essential is to use a sufficiently large scale to minimise the effect of enlargement of the image due to structure of the emulsion. As a limiting case, an example that can be quoted is the definite elongation into a pear-shaped image of a star with components of magnitude 4 and 8 separated by 70" on a negative in which the scale was only 2.4mm per degree. This exposure was made with a quite ordinary telephoto lens of 135mm focal length and a linear aperture of 34mm on HP4 film.

Whilst photography cannot compete with the visual micrometer for absolute accuracy, it does have the advantage of providing a permanent record of the observation. There are a great number of double stars suitable for the camera, as a study of Webb's writings will reveal. Had Webb had the apparatus and materials that are available today he would certainly have tried his hand at photographing double stars, and it is hoped that the information in this chapter will prompt the reader to active work in this field.

7. BRIEF BIOGRAPHIES OF DOUBLE STAR OBSERVERS.

Robert Grant Aitken. (1864 - 1951)

Robert Grant Aitken was born in Jackson, California on December 31, 1864 and was educated at Oakland High School. After graduating from Williams College in 1887, Aitken became an instructor in mathematics at Livermore College. Between 1891 and 1895 he was Professor of Mathematics at the University of the Pacific at San Jose in California. At this time he became Assistant Astronomer at Lick Observatory and his earliest work in the field of double star observation consisted of measuring, with the 12-inch refractor, a list of pairs selected by Professor E.E.Barnard.

After the work carried out by the Struves, it had generally been supposed that there were few new double stars left to be discovered, but this idea was soon dispelled by Burnham who was continually adding to his catalogues. Aitken had realised however, that Burnahm's discoveries had not been the result of a systematic search as had those of the Struves. He was therefore convinced that given the excellent observing conditions at Lick, the great resolving power of the 36-inch O.G. and the advantage of systematic work, a large number of systems could still be discovered. It was in July 1899 that Aitken and W.J.Hussey (who arrived in January 1896) combined forces, and began to examine all the stars of the B.D. down to mag 9.0 (Hussey included those of mag 9.1) north of -22°. When Hussey left in 1905, Aitken continued alone, finishing in 1915.

In addition to his observational work, Aitken computed numerous orbits, and in 1920 after the death of Doolittle, took over the supervision of the extended Burnham card index. This was used in the compilation of his great double star catalogue, the ADS in 1932. It was in this year that Aitken was awarded the Gold Medal of the Royal Astronomical Society.

He was also the first President of Commission 26 of the IAU, and when succeeded by Ejnar Hertzsprung in 1928, he was elected Honorary President. In 1930 he became Director of Lick Observatory, and five years later, Emeritus Astronomer and Director at Berkeley, where he retired and spent his remaining years.

In addition to double star work he enjoyed making measures of cometary positions and computing their orbits as well as measuring the satellites of Mars, Jupiter, Saturn, Uranus and Neptune.

Dr. Paul Baize. (1901 -)

Dr. Baize, after graduating as a medical practitioner in 1924, has worked as a apecialist pediatrician during his long career. His astronomical work, although substantial by even professional standards, has been entirely amateur. He began serious double-star observing with a 10.8-cm refractor in 1925, equipped with a micrometer of his own making. Between 1925 and 1932 he made some 3,834 measures of double-stars which were published in the "Journal des Observateurs", some of them also being included in Aitken's "General Catalogue of Double Stars" of 1932.

In 1933, Dr. Baize began to use the 30.5-cm equatorial in the West Tower of Paris Observatory, with which he completed 11,332 measures by 1949. From 1949 to 1971 he was able to work with the 38-cm equatorial in the East Tower and, with this instrument, he produced an additional 8,878 measures.

In addition to his observational work, Dr, Baize has calculated the orbital elements of some 200 binary stars and has published a number of astronomical articles in "Annales d'Astrophysique", "Bulletin Astronomique" and "Journal des Observateurs". He has been a member of the Comité National Francais d'Astronomie since 1932, and served on Commission 26 of the IAU in 1936.

Sherburne Wesley Burnham. (1838 - 1921)

Sherburne Wesley Burnham was born in Thetford, Vermont on December 12, 1838. He began work as a shorthand reporter at first in New York and then later with the Federate Army in New Orleans. After the end of the Civil War, during which he first began to take an interest in astronomy, he continued in his profession and in 1892 became Clerk to the U.S. Circuit Court, a position which he held until 1910.

On a visit to London in 1861, Burnham purchased a 3-inch refractor which was designed more as a terrestial than an astronomical telescope. In 1866 he settled down in Chicago, his home being just a few hundred yards from the Dearborn Observatory, which had just been supplied with a new 18½-inch refractor by Alvan G. Clark and Sons. At this time Burnham was changing telescopes regularly, never being satisfied with his current acquisition.

In 1869 he chanced to meet Alvan G Clark in Chicago, and the meeting resulted in the ordering of the 6-inch refractor which was to win world renown for its owner. He stipulated that "its definition should be perfect and that it should do on double stars all that it was possible for any instrument of that aperture to do". Burnham himself was unable to explain why his interest should have focussed on double stars, but perhaps it was because he had amongst his small library a copy of Webb's "Celestial Objects for Common Telescopes". Burnham began his work by looking for new double stars, and found his first new pair, BU40, on April 27, 1870. When he had accrued a list of 81 probably new pairs he sent it off to the Monthly Notices of the RAS. This was followed about a year later by two further lists, although at this time Burnham did not have a micrometer. Subsequent contact with the great Italian observer Dembowski resulted in the latter supplying measures of Burnham's early pairs.

In 1874 Burnham was able to use the 9.4-inch refractor of the Dartmouth College Observatory on 10 nights whilst spending a summer vacation in New Hampshire. On this trip he also observed with the great 26-inch refractor at Washington on one night, and then added 14 new pairs to his growing list. From then on Burnham was able to use many of the great American telescopes, culminating in the 36-inch at Lick and the 40-inch at Yerkes, and apart from the period 1888 - 1892, when he was officially on the staff of Lick Observatory, he remained an amateur observer following his profession during the day and at weekends making the long trip out to Yerkes.

Early in his career, the lack of really useful astronomical literature on double stars prompted him to make notes at every opportunity of all the measures of the known pairs - a task which later expanded to become the General Catalogue of Double Stars, published in two volumes in 1906. This magnificent work contains a valuable collection of 10,000 measures of the more neglected doubles as well as generous measures of the binaries known at that time.

He continued to observe until 1913 and during his long career discovered 1,274 double stars, many of which are in orbital motion, and some of which are amongst the binaries of shortest known period. It is a tribute to Burnham's superb eyesight that some of his most difficult discoveries were first suspected with the 6-inch refractor and later confirmed with a larger telescope in which they were still very difficult objects.

Burnham's fundamental contribution to astronomy did not go unnoticed, and in 1894 he received the Gold Medal of the Royal Astronomical Society, followed ten years later by the Lalande prize of the Paris Academy of Sciences. He also received honorary degrees from Yale and Northwestern Universities.

Rev. William Rutter Dawes. (1799 - 1868)

A qualified medical practitioner and an Independant Congregational Minister, W.R. Dawes had been interested in astronomy from his youth. He did most of his double-star observing between 1841 and 1860 at Cranbrook in Kent, using a 6-inch Merz refractor. Here, he produced a "Catalogue of Micrometrical Measures of Double Stars" containing data on some 121 pairs.

Later, he moved to Haddenham, Bucks., where he used an 8-inch Cooke refractor in 1865 to make further double-star measures. The Lick Index of Double Stars contains 15 pairs attributed to Dawes. (See WSQJ Vol.4, No.2, Jan 1972, p.10).

Dawes, who is often remembered as "the eagle-eyed", as Sir George Airy called him, published a catalogue of double stars in 1867.

Biographies.

Baron Ercole Dembowski. (1815? - 1881)

Dembowski's astronomical work was almost exclusively devoted to the accurate measurement of double and multiple stars. He began observing in 1852 at his private observatory in Naples, using a 5-inch dialyte by Plosse, and in 1857 he published accurate measurements of 127 double and triple stars, taken from F.G.W. Struve's Dorpat Catalogue. Dembowski produced further revisions of the Dorpat Catalogue in 1860, 1864 and 1866.

In 1870 Dembowski moved to Milan where he set up a new observatory at Cassano Maganzo, equipped with a 7-inch Merz refractor. With this instrument he made a complete and thorough revision of the Dorpat Catalogue, the final results being published (after his death) in 1883 at Rome in the two volumes of his well-known "Misure Micrometrice di Stelle Doppie e Multiple".

Dembowski received the Gold Medal of the R.A.S. in 1874 for his double-star work which, from the beginning of his astronomical career, was celebrated for its thoroughness and accuracy.

Eric Doolittle. (1869 - 1920)

The son of a Professor of Mathematics and Astronomy at Lehigh University, Pennsylvania, Eric Doolittle was born on July 26, 1869 in Ontario, Indiana. After schooling in Bethleham, Pennsylvania he went on to attend Lehigh University, graduating as a civil engineer in 1891. Following Aitken and Hussey, he became instructor in mathematics at Lehigh and the University of Iowa, and in 1894 commenced postgraduate study at the University of Chicago. During this period of two years, he did research on the "Secular Variations of the Orbits of the Four Inner Planets" which was published as a memoir in 1912.

In 1896 Doolittle returned to the University of Pennsylvania to become instructor in mathematics, and in addition, assistant to his father, Charles, who was Director of the newly founded Flower Observatory. The main instrument was a fine 18-inch Brashear refractor of 30 feet focal length. For most of the rest of his life, Doolittle engaged in double star measurement, and between 1901 and 1914 published four large volumes of measures. The first was "Measures of 900 Double and Multiple Stars" followed by "Measures of 1,066 Double and Multiple Stars" in 1905, "Catalogue and Re-measurement of the 648 Hough Double Stars" in 1907 and "Measures of 1,954 Double Stars" in 1914. In fact several thousand measures were made after 1914 but remained unpublished in Doolittle's lifetime.

In 1913 S.W. Burnham, on his retirement, entrusted the additional data for an extension of his General Catalogue (1906) to Doolittle, who kept methodical full records of all the published measures of double stars. Doolittle did not live to see the completion of this work, which was carried on by his friend Robert Aitken, but died on Sept 21, 1920.

Biographies.

Maurice Victor Duruy. (1894 -)

Maurice Duruy, a qualified Mining Engineer of the Ecole Polytechnique and School of Mines in Paris, graduated from the University of Nancy. His whole career was spent working for the French Government as an Engineer and then an Inspector of security in mines, which involved travelling abroad, notably to Britain and South Africa. At the same time he undertook a teaching post in applied mechanics at the Mining Institute of the University of Nancy, training future mining engineers. He later became Director of the Institute, and went on to become Administrator of three French Mining Schools. He retired in 1965.

His astronomical career had started much earlier. The first visual observations date from 1907 and show Saturn below Pegasus where it was to return in 1937 and 1966. In 1910, at the age of sixteen, his first telescope (a 3-inch refractor) was acquired, and the first observing notes made. Amongst these notes (which have been kept to this day) are the passage of the Earth through the tail of Halley's comet on May 18, 1910 and the total eclipse of the Sun visible in Paris on April 17, 1912. A 5-inch equatorial, obtained in 1912, enabled a serious study of celestial objects to be made. This instrument remained in Brittany, and for Paris M. Duruy used a 5-inch reflector in 1921 which was soon replaced by an 8-inch Cassegrain at Nancy. Up until 1928 all the observations were purely descriptive.

The second main period in his astronomical career started in 1934 with the purchase of an 18-cm objective. The programme of work included double and variable stars along with Mars, Jupiter and Saturn, and observations were published in the "Journal des Observateurs". The success of the refractor persuaded him to order a 27.5-cm objective from the well-known optician André Couder. This was finished at the end of 1936. Publication of observations became annual until the war, when the observatory was destroyed. The objective which had been used for just over 2 years was saved, but could not be used because of its long focal length. The war caused M. Duruy to be called to Lyon in June 1941 where he received a warm welcome from the Official Observatory there. The 16.2-cm and then the 32.5-cm refractors were put at his disposal, and he worked there until 1946. Difficulties in finding accommodation did not allow him to take up observing again until 1962 in the Montlhery region near Paris. A 26-cm reflector gave good results with double stars in the calm atmosphere. In 1964 a 40-cm equatorial reflector was mounted under cover with the same programme of doubles and variables being carried out.

In 1966 M. Duruy retired to central France where, in 1969, a 60-cm equatorial was added to the 40-cm instrument. These two telescopes are now used together, the 60-cm working in an observing hut which allows it to see the meridian whilst the 40-cm altazimuth is out in the open. The site of the Observatory between mountains (at 6,100 ft) and the warm sea gives troublesome atmospheric turbulence. Variables now take precedence over double stars which can only be observed occasionally.

Before 1940 M. Duruy published principally articles on the errors of double star measures. After 1950, he made many observations of the satellites of the larger planets and also contributed to the work in variable stars for the bulletin of the A.F.O.E.V. which he directed.

M. Duruy still contributes to the W.S.Q.J. and in the Catalogue to this Handbook some 3,499 measures of 420 pairs are due to him.

Rev. Thomas Henry Espinall Compton Espin. (1858 - 1934)

Educated at Haileybury and Exeter College, Oxford, Espin took Holy Orders, and from 1888 until his death was Vicar of Tow Law in County Durham. He became interested in astronomy after seeing Coggia's comet in 1874, and at Oxford he was allowed to use the 13-inch De la Rue telescope.

The first part of his work for which he became well-known was his observation and cataloguing of red stars of which he had listed some 3,800 between 1885 and 1899.

In 1900 Espin started observing double stars with a 17¼-inch Calver reflector - the first to use an instrument of this type for double-star work since the Herschels. Most of his discoveries, therefore, were confined to comparatively wide pairs, and these were found by re-examining the stars of the B.D. Catalogue. The total number of his own discoveries in this field was, according to his own numbering, 2,575.

One notable observation of a different kind was his discovery of Nova Lacertae on 1910 Dec 30.

Espin was very friendly with Webb and assisted him in producing some of the earlier editions of "Celestial Objects". After Webb's death, it was Espin who published the 5th and 6th editions of this famous handbook.

Espin's work was continued by his assistant and colleague W. Milburn, who also used the 17¼-inch Calver reflector to discover another 1,051 pairs.

William Stephen Finsen. (1905 -)

Although born in Johannesburg Finsen is of Icelandic descent, being the nephew of Niels R. Finsen, winner of the 1903 Nobel Prize for Medicine, and founder of the Finsen Institute in Copenhagen.

His early years were spent in Denmark, but in 1912 he returned to South Africa, and after studying at the King Edward VII School in Johannesburg, Finsen went on to obtain a B. Sc. at the University of

South Africa in 1930. Further degrees of B.Sc. Honours and M.Sc. were awarded in 1936 and 1937 by the University of Witwatersrand, and in 1951 he received a D.Sc. from the University of Cape Town.

Finsen assisted van den Bos in the Union Observatory survey of southern double stars. Later, he specialised in observational and theoretical work on double stars, but was for some years in charge of the Time Department at the Observatory. In 1933 he began tests with the Anderson double star interferometer, and in 1950, designed and constructed an eyepiece interferometer. With this instrument attached to the 26½-inch refractor at Johannesburg more than 13,000 examinations of 8,117 stars between declinations -75° and $+20^{o}$ were made. As a result 73 new pairs were found, 11 of which have orbital periods ranging from 21 years down to 2.65 years. In addition, 6,000 measures of pairs too close for the micrometer were made.

Finsen also discovered the "splitting" of Nova Pictoris in 1928, studied the rotation of Eros visually in 1931, and during the oppositions of Mars in 1954 and 1956, took many fine photographs which were extensively reproduced in journals and books.

He became Assistant Director in 1941, Director in 1957, and retired in 1965. However, he continued to carry out private research after that time. He is the author of some 135 papers and reports including three editions of a catalogue of visual binary orbits (the last in 1970 in collaboration with C.E. Worley of USNO).

Sir Frederick William Herschel. (1738 - 1822)

As in many other branches of astronomy, in double-star observation, William Herschel was a magnificent pioneer. His first observations, of close unequal pairs, were designed to obtain a value for stellar parallax and, beginning in 1799, he soon produced a list of 269 pairs, only 42 of which had previously been observed as double. Eventually Herschel, after making numerous accurate measurements of distance and position angle, was forced to the conclusion that the changes he detected were due not to parallactic shift but to actual orbital movement of the pairs.

This extremely important conclusion was revealed in his Phil. Trans. paper of 1803, and this date represents the real beginning of double-star astronomy. It should be remembered, too, that at the time, this evidence of true binary orbital motion showed that the principles of Newtonian gravitational theory could be extended well beyond the solar system, and strongly supported the view that this was a truly universal law.

In his double-star measurements, as in all his work, William Herschel displayed his genius: where equipment and instruments, as he found them, were inadequate for the task in hand, he improved them and used them with consumate skill; when the evidence his observations revealed was at variance with the original hypothesis, he accepted it without question, and immediately drew the correct conclusion and revealed a new and vitally important aspect of the heavens.

Sir John Herschel. (1792 - 1871)

When John Herschel was persuaded by his father to give up his law studies and devote himself to astronomy, he began his work in 1816 on purely double-star observations. His object was to re-measure some 364 of his father's discoveries in order to accrue more evidence in support of the binary orbital motion which had been adduced earlier.

John Herschel then began a fruitful collaboration with James South (later Sir James South) at Southwark which resulted in the publication of a catalogue of 380 new pairs. For this work Herschel and South were jointly awarded the Gold Medal of the R.A.S. in 1826.

John Herschel's expedition to the Cape of Good Hope (1834 - 38) was one of the most fruitful in the history of astronomy and although in these southern observations, double-stars were only "of subordinate interest", he nevertheless managed to catalogue some 1,202 pairs.

After his return from the Cape, he devoted his time to publishing the results of his visit, which occupied him until 1847. He made few astronomical observations after this time but his pioneer observations with South were of great importance in double-star astronomy.

William Joseph Hussey. (1862 - 1926)

Hussey graduated at Ann Arbor University in 1889. Between 1889 and 1891 he was, like Aitken, an instructor in mathematics, and the following year became Assistant Director of the Detroit Observatory of the University of Michigan, until Professor Asaph Hall took up his appointment as Director.

Hussey then moved on to the Layland Stanford Junior University at Palo Alto where he became Professor of Astronomy. Whilst here he would make frequent visits to Lick, and in 1896 joined the staff as Astronomer. In the years 1898 and 1899 he re-observed all of the Pulkova double stars of Otto Struve, the results appearing in Publications of the Lick Observatory, Vol. V, 1901. This very valuable work also contains a complete collection of every other measure published of these pairs, together with their respective references.

During his co-operative programme with Aitken he found 1,327 pairs in six years, for which he was awarded the Lalande Gold Medal of the Paris Academy of Sciences.

In 1905 he returned to Ann Arbor to become Professor of Astronomy and Director of the Detroit Observatory, and during this time designed a $37\frac{1}{2}$-inch reflector which was subsequently built almost entirely in the workshops of the University. In 1911 he became Director of the Observatory of the University of La Plata in Argentina whilst retaining his post at Ann Arbor. He made four trips out to Argentina, and using the 17-inch refractor there, added a further 323 pairs to his catalogue of discoveries.

Hussey's greatest ambition was to see the establishment of an observatory in the Southern Hemisphere for the discovery and measurement of double stars. His early college friendship with R.P. Lamont led to the financial support of the latter for a 24-inch refractor. The interruption of the war delayed the delivery of the blanks for the objective but in 1923 two 27-inch blanks were obtained from Jena. In the same year Hussey journeyed to Bloemfontein to look for a suitable site for a new observatory. In October 1926 while in London on his way to South Africa to supervise the erection of the telescope, he died suddenly.

Although, tragically, he did not live to see the completion of the project, the telescope was soon in working order under the supervision of Professor R.A. Rossiter, and it has since fully justified Hussey's expectations.

Robert Thorburn Ayton Innes. (1861 - 1933)

Innes was born in Edinburgh and went to school in Dublin, but all his astronomical work was carried out in the Southern Hemisphere. Beginning purely as an amateur, he soon became well-known in Australia for his discovery and measurement of double-stars, besides gaining a reputation for his ability in mathematical astronomy, including celestial mechanics.

In 1896 he moved to South Africa where he became secretary to Sir David Gill at the Cape Observatory. Here he used the 7-inch and 18-inch equatorials for the observation of both double and variable stars, and in 1899, compiled a "Reference Catalogue of Southern Double Stars".

In 1903, Innes became Director of the Transvaal Observatory at Johannesburg: here, in addition to photographic work with the twin 6-inch and 7-inch Franklin-Adams refractors, he made a thorough and systematic survey of the southern skies with the $26\frac{1}{2}$-inch refractor for the discovery of new double stars. Together with B.H. Dawson of La Plata Observatory and W.H. van den Bos, Innes compiled and published a "Southern Double-Star Catalogue" in 1927. In this important work, Innes contributed 1,613 new pairs, van den Bos more than 2,000 and W.S. Finsen more than 300.

R.T.A. Innes made many other contributions to astronomy, including the confirmation (from a reduction of all known Transits of Mercury) that the rotation of the Earth is not constant, and that Proxima Centauri is our nearest stellar neighbour.

Biographies.

Robert Jonckheere. (1889 - 1974)

Jonckheere began his long career as a double-star observer in 1905 using a 3-inch refractor. This instrument was followed by a 4-inch and 5-inch, and in 1907, by a 9-inch equatorial refractor fitted with a micrometer. This work was carried out at a private observatory on the roof of a house in Roubaix but soon Jonckheere decided that he needed more formal instruction in astronomy and he took a course at Strasbourg Observatory where he used a 6-inch equatorial. With this instrument he discovered 40 new pairs which were included in a list he published in the Bulletin Astronomique in December 1908.

Jonckheere now set about the task of setting up a Belgian Observatory and this was successfully established at Hem near Lille and eventually was attached to Lille University. Double-star observation was a major programme at Lille, but in 1914 the country was over-run by World War I and Jonckheere took refuge in England. Here,he was able to use the 28-inch refractor at Greenwich Observatory and by the end of 1916 he had compiled a catalogue.

In 1919, Jonckheere was able to return to Lille Observatory to continue his double-star work without further interruption. In 1930 he retired and moved to central France: his retirement, however, was only nominal, for during the next 30 years, he was able to use various instruments at the Observatories of Marseilles, Nice, Toulouse and Strasbourg.

Eventually, in 1962 he published his major work, the "General Catalogue" which contains, among other observations, no less than 3,355 new double and multiple stars discovered by him since the beginning of his career in 1906.

Although his life-long interest lay in double-star observation, Jonckheere made many observations of planetary nebulae, red stars and novae. His published papers amount to more than 40, and represent the devotion to astronomy of a long and fruitful life. He died on June 24th 1974.

Thomas Lewis. (1856 - 1927)

Lewis joined the staff of the Royal Observatory early in 1881. Initially his work involved the reduction of zenith distance observations made with the Transit Circle (8.1-inch aperture, f.l. = 11' 7") but later in 1881 became Superintendant of the Time Department, a post which he occupied until his retirement in 1917.

Lewis spent several years observing with the Transit Circle, the altazimuth (3.75-inch aperture for lunar observations) and the Small Equatorial. In 1893 he started to measure double stars with the Great Equatorial, a refractor with a 12.8-inch Merz object glass of 17' 10" f.l. The new 28-inch refractor was delivered in the same year and Lewis spent most of his observing hours using this telescope for double star measures.

Biographies.

He concentrated on the pairs of F.G.W. Struve and is most remembered for his great volume, "Measures of the Double Stars contained in the Mensurae Micrometricae of F.G.W. Struve." which was published in 1906. This earned him the Lalande prize of the Paris Academy of Sciences the following year.

He retired in September 1917 and lived at Wivenhoe in Essex until his death in June 1927.

Fr. Angelo Secchi. (1818 - 1878)

Although best-known for his pioneer study of stellar spectra and their classification into four types, Secchi made a significant contribution to double-star work. Using the fine 9.6-inch Merz refractor of the Observatory of the Collegio Romano (where he was Director), Secchi observed double stars between 1854 and 1859, publishing the results in his "Catalogo do 1,321 Stelle Doppie..." of 1860. In this work, Secchi also attempted a statistical analysis of the proportion of binary pairs to the total in F.G.W. Struves' "Mensurae Micrometricae" and related to the degree of separation of the pairs. The Lick Index of Double Stars (IDS) lists only four pairs against Secchi. (See W.S.Q.J. Vol.5 No.1 July 1972 p12).

Admiral William Henry Smyth. (1788 - 1865)

The son of an American loyalist emigré, he started his career in the Royal Navy as a boy. He obtained a commission from the lower deck and commanded a brigantine in the Mediterranean during the Napoleonic wars.

Smyth owed the intense interest in astronomy of his later years to a meeting with Piazzi in 1817 at Palermo Observatory. His first astronomical telescope was a fine 5.9-inch refractor by Tulley which he set up in his small observatory at Bedford in 1830. He took a special interest in double-star observation. By 1839 he had catalogued many celestial objects, including some 350 double and multiple stars, some of the binaries being measured annually. After this date Smyth sold his equipment which was installed in a new observatory belonging to a Dr. Lee-Smith at Hartwell near Aylesbury.

Smyth retained access to the observatory at Hartwell and by 1843 had made measures, accurate for his day, of some 680 pairs. In the following year 1844, Smyth published his "Cycle of Celestial Objects", the work for which he is generally remembered and for which he was awarded the Gold Medal of the R.A.S. in 1845. In 1860 he published another book, "Speculum Hartwellianum", which, although not as popular as his "Cycle", contains a large number of double-star measures.

Biographies.

Sir James South. 1785 - 1867)

South was a qualified surgeon who became interested in astronomy and, after acquiring a fortune by marriage, spent his whole time and considerable sums of money on his chosen hobby. He began double-star observations with a 5-inch refractor at his observatory in Blackman Street, Borough, London. He became a founder-member of the newly-formed Astronomical Society of London and through his influence, obtained the Royal recognition of the Society.

His friendship with John Herschel led to their collaboration in remeasuring William Herschel's doubles, and their joint catalogue of 380 pairs was presented to the Royal Society in 1824.

Immediately after this, South moved to Passy, near Paris where he began a further series of double-star measures, but by 1826 he had returned to England, establishing a very well-equipped observatory at Campden Hill, Kensington. Here, South had ideas of making a comprehensive double-star survey and he spent a great deal of money in obtaining a $11\frac{3}{4}$-inch Cauchoix objective and getting it mounted by Troughton. He soon became dissatisfied with the telescope, declaring it useless for his purpose and refusing to pay Troughton's bill. Many law-suits followed - much of the trouble being due to South's intemperate behaviour - and he abandoned his double-star ambitions and destroyed his equipment. The Cauchoix objective was saved however and eventually found its way to Dunsink Observatory where it is still in use.

Friedrich Georg Wilhelm Struve. (1793 - 1864)

Otto Wilhelm Struve. (1819 - 1905)

F.G.W. Struve was one of the greatest double-star observers and, after the Herschels, a pioneer in the field. He began his major astronomical work at Dorpat in 1812. Although his equipment was meagre, he managed to produce his "Catalogus 795 Stellarum duplicum" in 1822. In 1824 Struve obtained a superb telescope, a 9.6-inch Fraunhofer refractor, then the largest in the world, equipped with an equatorial mounting and driving clock. With this instrument he began his monumental survey of double stars down to 15° South. In a little over two years, Struve, with his assistants, Preuss and Knorre, completed no less than 10,448 measures of some 3,112 pairs and multiple stars of which 2,343 were new discoveries.

The results of this great work were published in 1837 as "Mensurae Micrometricae Stellarum duplicum et multiplicum" a catalogue which was to prove the foundation for many subsequent observers to build upon.

In 1839 Struve became Director of the newly-established Pulkova Observatory where the principal instruments were a 15-inch refractor and a 7.6-inch heliometer. The main programme consisted of a comprehensive general survey of Northern Hemisphere stars down to mag. 7 and a positional survey of equatorial stars. In this work he was assisted by his son Otto Struve who also continued with double-star observations, eventually publishing a catalogue of 547 new pairs (known as the Pulkova stars) in 1850.

In many subsequent catalogues, F.G.W. Struve's doubles are prefixed with a capital sigma (Σ) and those of Otto by ($O\Sigma$).

George Van Biesbroeck. (1880 - 1974)

Among double-star observers, George Van Biesbroeck must be considered as one of the greatest. He was an active observer for more than 70 years and the major part of his work was devoted to double-star astronomy. Although he started his working life as an engineer, graduating at the University of Ghent in 1902, he soon transferred his energies to astronomy, beginning as a volunteer observer at the Belgian Royal Observatory at Uccle. Using a 15-inch refractor, his careful micrometer measurements of close doubles earned him immediate recognition.

In 1906, he obtained a Fellowship at Heidelberg University and continued his double-star measures with the 12-inch refractor until 1908, when he returned to Uccle. Like many other European astronomers, Van Biesbroeck found life and work greatly hampered by World War I but in 1915 he was fortunate in obtaining a Visiting Fellowship at Yerkes Observatory and in the following year he and his family emigrated to Wisconsin.

At Yerkes Observatory, he found in the famous 40-inch refractor, an ideal instrument for the study of doubles that were both close and faint, and in his first programme he re-observed most of the 1,200 or so doubles which W.J. Hussey had discovered earlier at Lick Observatory. As might be expected, many of these close pairs showed fairly rapid change since Hussey's time, and Van Biesbroeck was able to compute reliable orbits for many of them.

During the mid 1930's Van Biesbroeck took part in the engineering work for the new 82-inch reflector at McDonald Observatory, and after its completion soon showed that a good reflecting telescope was just as effective as a refractor for double-star measures.

Although he discovered some new doubles during his long career, Van Biesbroeck concentrated his attention on the continuing observation and precise measurement of close doubles which were already known, thus providing reliable data from which orbital elements could be obtained.

In 1945, Van Biesbroeck, now aged 65, officially "retired" but in fact he continued observing: he took part in eclipse expeditions; to Brazil in 1947 and the Sudan in 1952, and in 1964 worked with the large reflectors of Kitt Peak and Steward Observatories.

Besides his outstanding double-star work, he made many observations of comets, minor planets (two of the former and one of the latter bear his name) and other objects. One of the interests linked to his double-star work was the search for companions to stars of large proper motion. In 1944 he discovered a faint companion to BD +4° 4048."Van Biesbroeck's Star",as it is now called, has an absolute magnitude of +19 and is still the least luminous star known.

Van Biesbroeck died on February 23, 1974: he lived a long and fruitful life during which his 36,000 or so double-star measures of the highest quality have put him among the foremost double-star observers of all time.

Willem Hendrik Van Den Bos. (1896 - 1974)

Dr. Van Den Bos was born in Rotterdam on September 25th, 1896. He was attracted to astronomy and in particular the study of double stars at an early age. In 1913 he commenced study at the University of Lieden, and whilst there was taught by such eminent men as Ejnar Hertzsprung and Willem de Sitter. In 1925 Van Den Bos received his doctorate in Physics and Mathematics from Leiden, and in August of that year arrived at the Union (later Republic) Observatory at Johannesburg in South Africa as a guest observer. He played a major part in the systematic search for new double stars in the Southern Hemisphere, which continued the work of Aitken and Hussey at Lick.

In 1930 Dr Van Den Bos joined the staff of the Union Observatory and remained there until his retirement in 1956. However he continued his researches in a private capacity until ill health forced him to stop in 1966, at which point he had made nearly 72,000 visual measures, discovered some 2,900 new pairs and computed 150 orbits. In addition to the 26½ and 9-inch refractors, Dr Van Den Bos used the 27-inch at the Lamont Hussey Observatory, the 24 and 14-inch at Bosscha, Java, the 36-inch at Lick, the 40-inch at Yerkes and the 10½-inch at Leiden (all refractors) as well as the 36-inch and 82-inch reflectors at the McDonald Observatory in Texas.

He was a visual observer of great skill and could make accurate measures with amazing rapidity, sometimes completing as many as twenty in an hour of favourable viewing conditions. This great double-star observer died on March 30th 1974.

PART TWO.

8. A CATALOGUE OF DOUBLE STARS.

INTRODUCTION.

The measures of 522 double, triple and quadruple stars in this catalogue were made by two observers: M.V. Duruy in France and R.W. Argyle in England. With the exception of two short lists of measures in The Astronomer, 1973 Jan and The Webb Society Quarterly Journal, 1972 Jul and a small number of measures for 1942/43 in Journal des Observateurs (to which additional unpublished observations have been added), all the measures included in this catalogue are previously unpublished.

The distribution of the measures is as follows:

Duruy: 3,499 measures of 420 systems.

Argyle: 377 measures of 170 systems.

These measures include 16 triple and 3 quadruple systems, the remainder being double. The distribution with separation is as follows:

Separation.	MVD	RWA	TOTAL
Single	41	-	41
Less than 2"	2,200	106	2,306
2" - 5"	729	57	786
5" - 10"	312	56	368
10" - 30"	154	102	256
More than 30"	63	56	119
Totals	3,499	377	3,876

The measures of Duruy cover the period 1941 - 1975 but they are not continuous, being divided into two distinct periods.

(a) 1941 - 1945.

During this time the 32.5-cm Coudé refractor of the Observatory of Lyon at St Genis Laval was used. The micrometer was a filar with a screw constant of 7".43 per revolution.

(b) 1963 - 1975.

Reflectors of 26-cm, 40-cm and 60-cm aperture were used and observations were made from two different sites.

(i) 26-cm Reflector.

This telescope was situated at Montlhery near Paris (altitude 80 metres) where Duruy reported that the seeing was such that measures could be made on more than 95% of clear nights. Between 1962 and 1964, measures of distance were made using the binocular micrometer which utilises a pair of artificial stars, whilst a wire in the eyepiece and a graduated circle were used for the measurement of position angle. From early 1964 to early 1965, the grating micrometer was used for pairs wider than 1".0 whilst below this separation the binocular micrometer was used as before.

(ii) 40-cm Reflector.

This was set up at Montlhery, the first measures being made in early 1965. Again the binocular micrometer was used for the great majority of measures. In mid-1966 M. Duruy moved to Le Rouret which is 25 kilometres west of Nice. Mountains rising to 2,000 m metres are 16 kilometres north and the Mediterranean is the same distance south of the Observatory. The Mistral frequently reduces the seeing to 5", making the measurement of close pairs impossible. The 40-cm and 26-cm were both equatorial.

(iii) 60-cm Reflector.

The 40-cm was transferred to an altazimuth mounting and the original equatorial, after reinforcement, was uset to carry the 60-cm. The telescope is housed in an observatory with a divided roof which opens to the west and east. At low declinations the 60-cm can see objects from 1 hour west to 2 hours east, whilst at high declinations, the limits are 2 hours west and 3 hours east.

Duruy notes that the poor seeing often prohibits the use of the telescope on double stars. A plane parallel window was added in 1970, and on nights of good seeing equally bright stars separated by only 0".20 can be measured.

Those measures by Argyle were made in two short periods in 1970 and 1972 - 1973. These were made using:

(i) The 28-inch Refractor at the R.G.O.

These measures were made between May and September 1970. For pairs below 4", the comparison image micrometer (C.I.M.) was used for separation exclusively, whilst the filar was used for all measures of position angle and those separations less than 4". The constant of the C.I.M. was 1mm = 0".0794 and that of the filar was 6".11 per revolution with a x2 Barlow lens fitted and 12".16 without.

(ii) 6-inch Cooke Refractor at the R.G.O.

This instrument was used between November 1972 and March 1973. A filar micrometer with a screw constant of 23".70 per revolution was used for all of these measures.

Double Star Catalogue.

EXPLANATORY NOTES.

The following abbreviations are used:

A	- R.G. Aitken	Hu	- W.J.Hussey
AC	- Alvan Clark	J	- R. Jonckheere
AGC	- Alvan G. Clark	Kru	- A. Kruger
Bu	- S.W. Burnham	Kui	- G.P. Kuiper
Cou	- P. Couteau	P	- Piazzi Smyth
D	- Baron E. Dembowski	SE	- Father A. Secchi
DA	- Rev. W.R. Dawes	SHJ	- Sir J. Herschel/Sir J. South
ES	- Rev. T.H.E.C. Espin	STF	- F.G.W. Struve
H	- Sir William Herschel	STF Ap	- ditto Appendix
HJ	- Sir John Herschel	STT	- O. Struve
HLD	- E.S. Holden	STT Ap	- ditto Appendix
HLM	- E. Holmes	WNC	- A. Winnecke
HO	- G.W. Hough		

The Right Ascension and Declination for 2000, the magnitudes, the original measures and the notes on comites were taken from the IDS. Optical comites fainter than 14.0 or whose magnitudes are not given have been excluded. The periods of binaries come from the works by Finsen & Worley (1970) and ephemeris positions for 1975.0 from Muller and Meyer (1969).

An "e" beside the measure of separation indicates that the pair was elongated. The number of nights on which measures were made is denoted by "N". If two figures are shown, the former refers to the number of measures of position angle and the latter to the number of measures of separation.

The codes in the extreme right hand column in the catalogue refer to the observer and telescope and are as follows:

D1	M.V. Duruy	32.5-cm refractor.	St Genis Laval
D2	M.V. Duruy	26-cm reflector.	Montlhéry.
D3	M.V. Duruy	40-cm reflector.	Montlhéry and Le Rouret.
D4	M.V. Duruy	60-cm reflector.	Le Rouret.
A1	R.W. Argyle	6-inch refractor.	Herstmonceux.
A2	R.W. Argyle	28-inch refractor.	Herstmonceux.

Double Star Catalogue.

In many of the wider pairs, the colour of each component has been estimated by members of the Double Star Section of the Webb Society. In order to make the notes more compact, the observed colours have been abbreviated thus:

w	- white.	c	- grey.
b	- blue.	r	- red.
o	- orange.	l	- lilac.
y	- yellow.	g	- green.
p	- purple.	a	- gold.

Intensities are denoted by the following:

P - pale. D - deep.

A tendency towards a particular colour is indicated by ysh (yellowish), bsh (bluish), etc.

Intermediate colours are represented by combinations of letters from the table above, viz: yw (yellow-white), bw (blue-white).

Colours were observed by the following: A. Tuck, D.A. Allen, B.Szentmartoni (Hungary), A. Fahy, A. Curran, J.E. Isles, I. Genner, A. Walker, E.R. Hancock, G. Gough, T. Northcott and the Director.

CATALOGUE OF DOUBLE STARS
(i) Measures

PAIR	RA 2000	DEC	MAGS	PA	SEP	DATE	N	OBS
STT 547	0005.4	4549N	9.4,9.4	167.9	5.36	67.61	4	D3
				169.3	5.80	73.59	3,1	D4
STF 3062	0006.2	5826N	6.4,7.5	150.8	0.51	43.68	5	D1
				167.7	0.6	45.19	1	D1
				177.7	0.62	45.86	6	D1
				255.4	1.27	65.61	6,5	D3
				260.8	1.28	67.63	3	D3
				271.4	1.31	72.53	3	D4
				272.0	1.40	73.54	3	D4
BU 255	0011.9	2825N	8.5,8.8	round		66.0	1	D3
BU 1026	0012.2	5337N	7.2,8.0	round		66.0	1	D3
STF 12	0014.9	0849N	6.1,7.7	147.5	11.6	72.9	1	A1
STT 4	0016.7	3629N	8.2,8.9	213.3	0.4	44.88	1	D1
				187.6	0.49	65.60	3	D3
STF 23	0017.5	0019N	7.9,10.2	278.3	3.75	42.90	2	D1
STF 24	0018.5	2608N	7.6,8.4	252.0	5.33	41.99	1	D1
STT 12	0031.8	5432N	5.5,5.8	177.7	0.56	65.70	6,5	D3
				179.1	0.52	70.64	3	A2
				176.2	0.55	73.62	2	D4
STF 40	0035.2	3650N	7.0,9.0	311.5	12.0	72.9	2	A1
STT 4 Ap	0036.8	3343N	4.4,8.9	173.4	36.1	72.9	3	A1
STF 1 Ap	0046.5	3057N	7.4,7.6	47.9	48.5	72.9	2	A1
STF 60	0048.8	5750N	3.6,7.2	284.1	9.66	43.97	2	D1
				286.1	10.00	45.92	1	D1
STF 60	0048.8	5750N	3.6,7.2	294.	10.22	62.9	1	D2
				297.6	11.05	64.6	2,3	D2
				299.9	11.25	66.0	3,5	D3
				300.7	11.64	73.62	5	D4
STF 61	0049.8	2743N	6.3,6.3	295.0	4.42	42.09	10	D1
				296.7	4.43	68.58	1,2	D3
BU 232	0050.4	5038N	8.5,9.0	225.0	0.72	65.59	3	D3
				244.	0.7	71.03	1	D4
STT 20	0054.6	1912N	6.1,7.2	238.7	0.42	65.90	5	D3
				238.8	0.47	67.66	5,4	D3
				237.0	0.5	71.02	2	D4
				226.7	0.48	73.71	4	D4
STF 73	0055.0	2338N	6.1,6.7	132.6	0.61	43.99	3	D1
				140.0	0.57	45.88	3	D1
				200.7	0.58	65.6	4	D3
				211.9	0.74	67.48	4	D3
				218.5	0.58	70.69	3	A2
				217.5	0.80	71.01	3	D4
				231.0	0.69	72.57	1	D4
				229.7	0.66	73.71	6	D4

PAIR	RA 2000	DEC	MAGS	PA	SEP	DATE	N	OBS
BU 500	0055.4	3039N	8.7,8.7	303.	0.4	71.03	1	D4
BU 1099	0056.8	6022N	6.0,6.7	round		65.6	1	D3
BU 302	0058.3	2125N	6.7,8.1	124.7	0.57	42.92	2	D1
				141.8	0.65	65.64	3	D3
				152.0	0.45	73.88	3	D4
STF 79	0100.0	4442N	6.0,6.8	194.3	7.93	44.62	3	D1
STT 21	0103.0	4722N	6.7,8.0	172.0	0.68	65.7	3	D3
STF 88 AB	0105.6	2128N	5.6,5.8	159.7	30.5	72.9	2	A1
AC			,11.2	72.9	108.2	72.9	1	A1
STT 515	0109.3	4715N	4.5,6.1	150.6	0.49	65.7	7,6	D3
				142.1	0.46	70.71	2	A2
BU 303	0109.6	2348N	7.3,7.5	289.0	0.67	67.65	2	D3
BU 235	0110.6	5101N	7.5,7.9	119.5	1.06	65.8	7	D3
A 655	0111.3	4113N	8.5,8.9	round		65.8	1	D3
STF 100	0113.7	0733N	5.6,6.5	63.7	23.0	72.9	2	A1
BU 1100	0114.7	6057N	8.3,8.3	43.	0.37	65.1	4	D3
				40.6	0.38	67.64	1	D3
STF 102	0117.8	4901N	7.1,8.3	284.5	0.45	65.63	4	D3
				278.8	0.42	71.03	3	D4
BU 4	0121.3	1132N	7.4,7.9	round		65.9	2	D3
STF 117CD	0126.0	6807N	9.4,10.0	73.2	2.98	42.93	2	D1
A 816	0135.7	7226N	8.5,8.6	307.5	0.57	42.97	2	D1
				310.1	0.62	43.99	2	D1
A 816	0135.7	7226N	8.5,8.6	308.2	0.53	73.89	4	D4
STF 138	0136.0	0739N	7.7,7.7	47.3	1.59	41.96	3	D1
				51.0	1.43	73.63	3	D4
HU 1030	0138.9	7644N	8.8,9.0	321.	0.6	43.97	1	D1
STF 162	0149.2	4754N	6.5,7.0	204.9	1.97	66.0	4,3	D3
STT 34	0149.9	8053N	7.9,8.1	258.5	0.35	66.0	1	D3
				259	0.31	67.65	1,2	D3
STF 174	0150.1	2217N	6.2,7.4	168.0	2.77	42.39	7	D1
				164.8	2.97	66.1	1	D3
				168.2	2.80	73.62	3	D4
HO 311	0151.2	2439N	7.6,7.8	200.5	0.3	71.02	1	D4
STF 180	0153.5	1918N	4.8,4.8	359.3	7.61	64.0	1,4	D2
				0.0	7.88	64.6	1,7	D2
				0.3	7.87	67.89	1,2	D3
STF 183	0155.1	2847N	7.8,8.5	round		65.6	2	D3
				round		73.63	1	D4
STF 186	0155.8	0151N	7.0,7.0	56.2	1.37	73.63	2	D4
STF 4 Ap	0156.1	3716N	5.8,6.0	298.3	196.6	73.0	2	A1
STF 182	0156.4	6116N	8.1,8.1	124.3	3.67	41.97	3	D1
STT 21 Ap	0158.1	2335N	4.9,7.7	47.1	38.5	72,9	2	A1
BU 513	0201.9	7054N	4.7,7.2	277.0	0.77	43.99	3	D1
				281.3	0.91	45.30	3	D1
STF 202	0202.0	0246N	4.3,5.2	294.2	2.38	63.0	3	D2
				291.7	1.95	65.6	3	D3

PAIR	2000 RA	DEC	MAGS	PA	SEP	DATE	N	OBS
STF 202	0202.0	0246N	4.3,5.2	291.7	1.97	67.05	7,10	D3
				288.4	1.82	73.58	3	D4
STF 205	0203.5	4223N	2.3,5.0	63.	9.96	64.5	1	D2
				64.7	9.63	70.7	1	A2
STT 38	"	"	5.1,8.3	101.1	0.4	42.27	2	D1
				97	0.4	44.02	1	D1
				115.	0.46	65.8	4	D3
				112.8	0.53	67.43	5	D3
				113.4	0.48	70.71	2	A2
STF 222	0210.8	3903N	6.0,6.7	35.9	16.8	72.9	2	A1
STF 227	0212.4	3018N	5.2,6.6	73.8	3.85	41.98	2	D1
				74.3	3.88	66.0	1,2	D3
STF 228	0214.1	4729N	6.4,7.3	210.2	0.47	44.02	4	D1
				252.6	0.81	65.64	3	D3
				254.6	0.87	67.47	3	D3
				260.9	1.06	73.60	4	D4
STF 234	0217.4	6121N	8.5,9.4	245.5	0.73	66.59	3	D3
STF 249	0221.6	4435N	7.2,9.2	195.4	2.23	42.98	2	D1
STF 257	0225.7	6133N	7.6,8.1	17.	0.30	65.64	1	D3
				24.	0.28	67.92	5	D3
				230.	0.25	71.35	3	D4
				45.6	0.22	73.64	1	D4
STF 271	0230.5	2515N	5.9,10.4	181.4	12.5	72.9	1	A1
STF 93	0231.6	8916N	2.1,9.1	216.4	17.8	73.2	3	A1
STT 42	0233.3	5218N	7.0,7.5	271.5	0.29	67.62	6	D3
STF 5 Ap	0237.0	2439N	6.6,7.4	275.6	38.3	73.0	2	A1
STF 285	0238.7	3325N	7.5,8.2	170.7	1.59	42.96	3	D1
				163.6	1.62	73.62	2	D4
STT 43	0240.7	2638N	8.3,9.9	30.4	1.14	44.02	1	D1
STF 299	0243.3	0314N	3.6,7.4	294.3	2.92	42.95	1	D1
STF 296	0244.2	4913N	4.2,10.0	301.8	18.36	42.99	1	D1
STF 305	0247.5	1922N	7.4,8.3	312.3	3.42	42.72	4	D1
				310.6	3.56	72.95	1	A1
STF 307	0250.6	5554N	3.9,7.9	300.9	28.4	73.0	2	A1
STF 314	0252.9	5300N	6.4,7.1	306.7	1.48	43.02	3	D1
				130.2	1.57	66.0	3	D3
BU 524	0253.7	3820N	5.6,6.7	344.8	0.25	73.64	3	D4
BU 525	0258.9	2137N	7.5,7.5	246.2	0.41	65.84	5	D3
				247.2	0.42	67.49	3	D3
				73.0	0.38	70.92	3	D4
				251.8	0.40	73.63	4	D4
STF 333	0259.2	2120N	5.2,5.5	205.3	1.57	65.9	2,1	D3
STF 331	0300.8	5221N	5.4,6.8	84.8	12.3	73.1	2	A1
STF 346	0305.5	2515N	6.9,6.9	round		44.02	1	D1
STF 360	0312.1	3712N	8.1,8.3	128.3	2.47	66.0	4	D3
STT 50	0312.5	7133N	8.5,8.5	175.0	1.27	66.0	3	D3
STF 367	0314.0	0044N	8.9,8.9	173.0	0.77	42.95	1	D1

PAIR	2000 RA	DEC	MAGS	PA	SEP	DATE	N	OBS
STF 367	0314.0	0044N	8.9,8.9	153.6	0.92	67.72	2	D3
STT 52	0317.5	6539N	6.9,7.5	80.7	0.42	66.0	6	D3
				63.2	0.47	73.63	2	D4
STT 53	0317.7	3838N	7.7,8.5	275.0	0.5	43.04	1	D1
				269.1	0.74	67.01	4	D3
				265.0	0.72	73.63	2	D4
STF 401	0331.4	2734N	6.4,6.9	269.5	11.1	73.1	2	A1
STF 396	0333.5	5846N	6.3,8.2	243.8	20.6	73.1	2	A1
STF 412	0334.5	2428N	6.6,6.7	46.0	0.4	43.16	1	D1
				13.8	0.50	67.06	4	D3
				8.2	0.62	73.64	1	D4
STF 400	0334.9	6002N	6.9,7.9	255.5	0.66	66.0	3	D3
				255.3	1.17	73.64	1	D4
BU 533	0335.6	3141N	7.6,7.6	43.7	0.87	42.96	2	D1
				46.5	1.11	67.73	2	D3
A 1535	0336.1	4221N	9.0,9.4	102.1	0.37	67.73	2	D3
STF 422	0336.8	0036N	6.2,8.4	259.0	6.74	43.10	3	D1
STT 37 Ap	0338.3	4449N	8.0,8.0	95.7	41.3	73.1	2	A1
STF 430AB	0340.5	0507N	6.8,9.8	57.4	26.3	73.1	2	A1
AC	"	"	,10.5	300.2	36.1	73.1	2	A1
STF 427	0340.6	2847N	7.3,8.1	207.5	6.77	43.10	3	D1
P III 97	0342.7	5959N	6.0,8.6	35.2	55.0	73.2	2	A1
STF 452	0348.3	1109N	5.0,10.1	58.6	9.5	73.0	2	A1
STT 65	0350.4	2536N	6.0,6.3	209.6	0.70	67.39	6	D3
				215.	0.6	71.22	3	D4
				208.5	0.66	73.65	3	D4
KUI 15	0351.9	0633N	6.3,6.4	35.	0.35	42.95	1	D1
				34.	0.42	66.0	1	D3
				28.5	0.6	67.59	3	D3
				38.8	0.5	73.77	1	D4
STT 66	0352.1	4048N	8.0,8.5	145.2	0.80	67.05	3	D3
STF 470	0354.3	0257S	5.0,6.3	347.	6.76	43.15	2	D1
STT 67	0357.2	6106N	5.3,8.5	49.8	1.41	66.0	2	D3
STF 479AB	0400.9	2313N	6.9,7.8	127.4	7.6	73.1	2	A1
AC	"	"	,9.0	241.4	57.8	73.1	2,1	A1
STT 531	0407.5	3805N	7.3,9.0	76.6	0.87	43.10	2	D1
STF 495	0407.6	1509N	6.0,8.8	221.0	3.59	73.1	2	A1
STF 460	0410.1	8042N	5.6,6.5	98.7	0.87	66.0	3	D3
STT 77	0415.9	3142N	8.2,8.2	72.8	0.65	43.04	1	D1
STT 44 Ap	0417.3	4613N	7.2,8.6	322.1	58.3	73.2	2	A1
STF 511	0417.9	5847N	7.4,7.9	124.1	0.36	67.07	3	D3
				130.	0.25	71.33	1	D4
STT 48 Ap	0420.4	2722N	5.1,8.5	254.0	49.6	73.1	2	A1
STF 528	0422.6	2538N	5.5,7.6	25.7	19.6	73.0	2	A1
STT 82	0422.8	1504N	7.3,9.3	206.2	1.18	43.10	3	D1
HU 304	0423.9	0928N	5.9,5.9	211.6	0.29	67.73	2	D3
STF 534	0424.1	2418N	6.2,8.6	289.9	29.1	73.1	2	A1

PAIR	2000 RA	DEC	MAGS	PA	SEP	DATE	N	OBS
STF 548	0428.9	3021N	6.5,8.5	36.5	14.7	73.1	2	A1
STF 550	0432.0	5355N	5.7,6.8	307.6	10.0	73.1	2	A1
STT 86	0436.6	1945N	8.2,8.2	40.6	0.46	43.11	3	D1
				24.5	0.44	67.09	2	D3
STF 567	0436.7	1929N	8.9,9.4	330.2	2.16	43.11	3	D1
STF 572	0438.5	2656N	7.3,7.3	196.5	3.90	42.52	6	D1
				193.6	4.24	73.16	2	A1
STF 11 Ap	0439.2	1555N	4.8,5.2	193.9	434.3	73.0	2	A1
STF 577	0442.3	3730N	8.6,8.6	46.0	1.27	43.13	5	D1
				33.4	1.08	67.21	5,4	D3
A 1013	0443.2	5932N	7.3,7.3	327.2	0.41	42.21	4	D1
BU 883	0451.2	1104N	7.8,7.8	240.	oval	66.0	1	D3
BU 552	0451.8	1339N	6.8,9.8	294.8	0.92	43.13	3	D1
STF 616	0459.3	3753N	5.1,8.1	357.5	5.64	42.92	5	D1
				0.7	5.35	73.0	2	A1
STF 623	0500.0	2719N	6.8,8.3	206.1	20.8	73.1	2	A1
STF 627	0500.5	0337N	6.6,7.0	260.3	21.7	73.0	2	A1
STF 630	0502.0	0137N	6.5,7.7	50.8	14.6	73.0	2	A1
STT 95	0505.0	1948N	7.0,7.6	314.7	0.92	44.15	2	D1
				306.8	0.97	66.2	3	D3
				307.8	1.02	67.01	3	D3
				305.6	0.75	73.77	1	D4
STT 98	0507.4	0830N	5.9,6.7	102.9	0.90	42.86	7	D1
				59.4	0.84	64.7	4	D2
				61.1	0.81	65.8	3	D3
STT 98	0507.4	0830N	5.9,6.7	57.6	0.86	67.10	3	D3
STF 635	0507.8	5459N	8.7,8.7	302.4	0.94	66.94	5	D3
STF 645	0509.8	2802N	6.1,8.6	28.9	11.8	73.0	2	A1
STF 654	0513.3	0252N	4.6,8.4	60.	7.00	43.10	1	D1
				64.0	6.73	73.1	3	A1
STT 517	0513.4	0158N	6.9,7.1	49.0	0.44	67.41	4	D3
STF 668	0514.5	0812S	0.1,9.3	203.7	10.1	73.0	3	A1
STF 653	0515.5	3241N	5.2,7.4	225.4	14.4	73.0	1	A1
STF 681	0520.7	4658N	6.7,8.7	181.6	23.2	73.0	2	A1
STF 634	0522.3	7918N	5.2,9.2	123.0	17.55	73.2	3	A1
STF 696	0522.8	0332N	5.0,7.1	29.5	32.0	73.0	2	A1
STF 701	0523.3	0826S	6.0,7.8	140.1	5.96	73.0	3	A1
WNC 2	0523.9	0053S	6.1,7.2	163.2	2.63	67.15	2	D3
A 847	"	"	8.0,8.1	151.	0.35	67.15	2	D3
DA 5	0524.4	0224S	3.8,4.8	79.1	1.48	43.14	3	D1
				81.8	1.42	67.13	4,3	D3
HO 226	0527.1	2736N	8.6,8.6	252.0	0.80	65.8	2	D3
				257.6	0.64	67.10	4	D3
STF 716	0529.2	2509N	5.8,6.6	203.6	4.75	43.13	5	D1
				205.7	4.96	73.2	3	A1
STF 728	0530.7	0556N	4.5,6.0	40.0	0.5	45.19	1	D1
				50.3	0.77	64.8	3	D2

PAIR	2000 RA	DEC	MAGS	PA	SEP	DATE	N	OBS
STF 728	0530.7	0556N	4.5,6.0	54.8	0.77	65.8	4	D3
				58.5	0.79	67.8	4	D3
STF 729	0531.2	0317N	5.8,7.1	25.7	1.84	43.11	4	D1
				26.0	1.90	64.7	3,4	D2
STF 14 Ap	0532.0	0018S	2.5,6.6	0.0	52.2	73.1	1	A1
STF 718	0532.3	4924N	7.5,7.5	73.5	7.5	73.2	2	A1
BU 1267	0535.1	3056N	8.8,8.8	201.9	0.6	65.76	2	D3
STF 738	0535.1	0956N	3.7,5.6	43.0	4.32	43.13	5	D1
				45.8	4.48	63.1	1,9	D2
STF 748CA	0535.3	0523S	5.4,6.8	311.4	13.2	73.1	2	A1
CB	"	"	,7.9	343.2	16.9	73.1	2	A1
CD	"	"	,6.8	61.1	13.4	73.1	2	A1
STF 752	0535.4	0555S	2.9,7.0	141.1	11.8	73.1	2	A1
STF 742	0536.4	2200N	7.2,7.8	267.1	3.75	42.20	5	D1
STF 749	0537.2	2656N	6.4,6.5	151.3	1.12	64.6	1	D2
BU 1240	0538.6	3030N	6.1,6.5	round		65.8	1	D3
STF 753	"	"	,8.0	268.3	12.2	73.0	1	A1
STT 112	0539.8	3758N	7.8,8.5	61.4	0.65	43.15	3	D1
				57.5	0.80	65.8	3	D3
				56.9	0.84	67.07	3	D3
STF 766	0540.3	1521N	7.0,8.2	274.2	10.1	73.1	2	A1
STF 774	0540.7	0157S	2.0,4.2	163.2	2.30	44.23	1	D1
				161.7	2.51	73.1	3	A1
BU 1007	0541.1	1632N	5.6,5.8	round		65.8	3	D3
STF 764	0541.4	2929N	6.8,7.4	13.3	26.0	73.0	2	A1
STT 115	0544.5	1503N	7.5,8.3	115.1	0.58	43.16	3	D1
				120.7	0.55	65.82	1	D3
BU 560	0547.4	2939N	8.0,8.0	141.4	1.28	43.10	1	D1
				131.8	0.41	65.76	1	D3
STF 795	0548.0	0627N	6.1,6.1	209.5	1.31	43.14	5	D1
				214.3	1.55	73.19	3	A1
STT 118	0548.4	2052N	6.1,7.6	311.2	0.4	65.8	2	D3
STF 3115	0549.1	6248N	6.5,7.6	358.7	0.86	67.08	3	D3
STF 802	0552.5	4008N	9.0,9.6	108.2	2.82	42.19	2	D1
A 1570	0557.6	4238N	9.9,10.4	350.3	1.20	67.29	2	D3
STT 545	0559.7	3714N	2.7,7.2	323.6	3.15	43.11	5	D1
				321.7	3.19	44.15	1	D1
				312.8	3.60	73.19	3	A1
STT 121	0605.3	7400N	7.6,8.8	round		67.2	1	D3
STF 855AB	0609.0	0230N	6.0,7.0	114.5	29.6	73.0	2	A1
AC	"	"	,8.9	106.6	88.5	73.0	1	A1
STF 134	0609.3	2426N	8.5,9.8	189.0	30.6	73.1	2	A1
STF 845	0611.7	4843N	6.1,6.8	356.4	7.7	73.1	2	A1
KUI 24	0614.4	1754N	6.5,6.5	138.3	0.57	65.80	5,6	D3
				138.	0.55	66.78	2	D3
				133.0	0.32	73.80	1	D4
STF 877	0614.7	1435N	7.3,7.8	263.9	5.80	43.17	5	D1

PAIR	2000 RA	DEC	MAGS	PA	SEP	DATE	N	OBS
BU 1008	0614.9	2230N	var,8.8	264.5	1.51	64.8	3	D2
				260.5	1.41	65.16	3	D3
STF 872	0615.6	3609N	6.8,7.8	216.2	12.0	73.1	2	A1
STF 881	0622.1	5923N	6.2,7.7	128.3	0.71	65.7	3	D3
STF 919BC	0628.9	0703S	5.7,6.2	286.9	2.80	42.20	1	D1
AC	"	"	4.7,6.2	123.2	10.12	42.20	1	D1
AB	"	"	4.7,5.7	131.1	7.01	42.20	1	D1
STF 924	0632.4	1747N	7.1,8.0	211.4	20.7	73.0	2	A1
STF 932	0634.3	1444N	8.1,8.2	319.5	2.00	43.17	3	D1
STF 928	0634.7	3832N	7.6,8.2	133.8	3.49	43.17	3	D1
STT 149	0636.5	2717N	7.0,9.5	round		65.8	1	D3
STT 152	0639.5	2816N	6.0,7.8	32.6	0.80	43.15	4	D1
STF 945	0640.4	4059N	7.3,8.2	295.2	0.55	66.0	5,3	D3
				114.7	0.60	66.97	4	D3
STF 946	0644.9	5927N	8.3,10.1	131.0	4.18	42.84	3	D1
AGC 1	0645.2	1641S	-1.4,8.5	72.	10.44	65.76	3,2	D3
STF 948AB	0646.3	5927N	5.3,6.2	97.7	1.65	43.15	7	D1
				95.4	1.68	45.19	1	D1
				89.8	1.76	65.70	3	D3
AC	"	"	5.3,7.2	305.1	8.63	43.15	7	D1
BC	"	"	6,2,7.2	303.1	10.22	43.15	7	D1
STT 78 Ap	0646.7	4335N	5.4,8.4	33.8	34.4	73.2	2	A1
STT 156	0647.4	1812N	6.7,7.2	274.7	0.61	43.52	4	D1
				78.8	0.58	65.79	2,3	D3
STT 156	0647.4	1812N	6.7,7.2	78.0	0.52	67.15	4	D3
STF 958	0648.3	5543N	6.3,6.3	258.8	4.73	42.89	6	D1
STF 963	0653.2	5928N	5.7,6.9	233.0	0.37	65.60	6	D3
STF 982	0654.6	1310N	4.8,7.1	149.2	6.80	73.07	3	A1
STF 981	0655.5	3010N	8.9,8.9	322.5	2.53	42.19	4	D1
				137.3	2.47	66.92	5	D3
STT 159	0657.3	5825N	4.8,5.9	38.4	0.91	65.64	4	D3
A 1575	0703.0	5403N	8.2,9.2	283.3	0.64	42.26	2	D1
STT 81 Ap	0704.1	2034N	var,8.0	348.4	98.7	73.0	2	A1
STF 1009	0705.6	5245N	6.9,7.0	149.5	3.71	67.55	1,2	D3
STF 1033	0714.8	5233N	7.7,8.3	275.7	1.53	66.0	2,3	D3
				277.9	1.44	67.14	3	D3
STT 170	0717.2	0919N	7.6,7.9	100.4	1.69	43.87	3	D1
				91.8	1.50	64.8	2	D2
				91.3	1.24	67.37	3	D3
STF 1066	0720.2	2159N	3.5,8.5	220.0	6.06	73.15	3	A1
STF 1083	0725.6	2030N	7.2,8.3	44.4	6.45	73.0	2	A1
STF 1093	0730.3	4959N	8.8,8.8	178.4	0.8	65.8	3,2	D3
STF 1108	0732.8	2253N	6.6,8.4	178.3	11.7	73.0	2	A1
STF 1110	0734.6	3153N	2.0,2.9	198.6	3.55	42.22	11,13	D1
				198.3	3.41	43.19	7	D1
				196.6	3.30	44.25	10	D1
				196.2	3.25	45.26	7	D1

PAIR	2000 RA	DEC	MAGS	PA	SEP	DATE	N	OBS
STF 1110	0734.6	3153N	2.0,2.9	156.0	1.95	63.14	11	D2
				153.5	1.97	64.76	8,7	D2
				150.8	1.93	65.76	10	D3
				144.2	1.89	67.29	6	D3
				119.8	2.07	73.11	3	A1
				117.5	2.0	73.77	2	D4
AB - C	"	"	1.7,8.8	163.3	71.8	73.1	2	A1
STT 175	0735.2	3058N	5.8,6.4	329.0	0.3	65.6	8,3	D3
STT 174	0735.9	4302N	6.5,8.1	86.9	2.23	73.1	3	A1
STF 1126	0740.1	0515N	6.4,6.7	154.7	1.10	43.88	3	D1
				159.0	1.08	65.82	3	D3
STT 177	0741.7	3726N	8.0,9.0	round		66.2	1	D3
STT 179	0744.5	2424N	3.7,8.2	240.4	6.99	73.1	3	A1
STF 1138	0745.5	1441S	6.1,6.8	340.1	16.6	73.1	2	A1
STF 1140	0748.5	1821N	7.9,9.6	275.4	6.3	73.0	2	A1
STT 186	0803.3	2616N	7.5,8.2	77.0	0.87	66.2	2	D3
BU 581	0804.3	1218N	8.7,8.7	284.1	0.51	43.16	3	D1
				178.6	0.37	66.2	4,3	D3
				186.1	0.34	67.25	4	D3
				193.6	0.37	68.2	3	D3
STF 1177	0805.6	2732N	6.6,7.5	350.5	3.44	42.97	8	D1
				168.7	3.50	65.8	3,2	D3
STF 1187	0809.5	3214N	7.1,8.0	32.5	2.48	44.20	3	D1
				27.9	2.78	65.81	2	D3
STF 1196	0812.2	1740N	5.6,6.0	76.4	0.86	42.55	5	D1
				68.6	0.81	43.17	7	D1
				64.2	0.89	44.23	4	D1
				56.3	0.92	45.29	4	D1
				176.6	1.06	63.10	4,2	D2
				344.8	1.14	65.81	7	D3
				342.3	1.02	67.20	3	D3
				340.2	1.11	67.84	3	D3
				335.7	1.09	68.23	4	D3
				329.2	1.1	70.94	4	D4
				316.9	1.00	73.22	4	D4
				313.4	1.00	73.79	4	D4
STF 1196	0812.2	1740N	5.6,6.0	93.5	5.51	63.10	4	D2
AB - C	"	"	5.0,6.3	84.8	5.69	73.79	4	D4
AC			5.6,6.3	95.2	6.38	42.22	4	D1
				95.9	6.28	43.37	4	D1
				95.9	6.20	44.23	4	D1
BC	"	"	6.0,6.3	98.7	5.41	42.22	4	D1
				100.5	5.43	43.57	4	D1
				100.6	5.43	44.23	4	D1
HU 1123	0814.8	3630N	8.9,9.2	156.2	0.44	43.19	2	D1
STF 1216	0821.4	0136S	6.9,7.6	242.5	0.42	65.8	1	D3
				264.6	0.48	67.19	2	D3

PAIR	2000 RA	DEC	MAGS	PA	SEP	DATE	N	OBS
STT 93 Ap	0824.6	4200N	6.2,8.7	174.0	82.6	73.1	2	A1
STF 1224	0826.7	2433N	7.1,7.6	48.2	5.67	42.24	4	D1
				48.8	5.80	73.2	3	A1
STF 1223	0826.8	2656N	6.3,6.3	217.3	4.98	42.24	5	D1
				217.0	5.39	73.1	3	A1
STF 1245	0835.8	0637N	6.0,7.2	26.0	10.6	73.0	2	A1
STF 1268	0846.7	2846N	4.2,6.6	307.4	30.6	73.0	2	A1
STF 1273	0846.8	0625N	3.6,7.8	260.8	3.01	43.25	6	D1
				277.5	3.25	73.05	3	A1
STF 1275	0851.5	5732N	8.4,8.4	198.5	2.05	43.28	2	D1
A 1584	0853.1	5458N	8.2,8.2	91.0	0.72	42.28	2	D1
STF 1291	0854.2	3034N	6.1,6.6	318.7	1.48	43.25	5	D1
KUI 37	0900.8	4148N	4.3,6.3	319.1	0.54	42.36	7	D1
				302.7	0.52	43.33	11	D1
				139.	0.4	73.77	1	D4
STF 1300	0901.3	1516N	9.3,9.4	192.4	5.12	43.30	2	D1
STF 1298	0901.4	3215N	5.9,8.0	136.7	4.43	73.1	2	A1
STF 1311	0907.5	2259N	6.9,7.3	199.5	7.75	73.1	2	A1
STF 1306	0910.4	6708N	5.0,8.2	45.4	1.77	43.26	5	D1
STF 1321	0914.9	5242N	8.1,8.1	84.7	17.68	73.19	3	A1
STF 3121	0918.0	2835N	8.1,8.1	55.5	0.3	43.21	3	D1
				29.0	0.62	73.22	2	D4
STF 1333	0918.5	3522N	6.4,6.7	49.1	1.77	43.27	4	D1
				49.0	1.77	63.0	4	D2
STF 1333	0918.5	3522N	6.4,6.7	48.4	1.82	73.37	5	D4
STF 1334	0918.9	3649N	3.9,6.6	228.5	2.44	63.5	2,4	D2
				231.0	2.56	67.02	2,5	D3
STF 1338	0921.0	3812N	6.6,6.8	243.0	1.18	73.39	4	D4
STF 1348	0924.4	0621N	7.5,7.6	139.4	1.98	43.30	3	D1
STF 1356	0928.5	0904N	5.9,6.7	151.0	0.84	43.62	7	D1
				358.5	0.53	73.23	4	D4
STF 1360	0930.5	1035N	8.3,8.6	242.9	13.8	73.1	2	A1
STF 1351	0931.5	6304N	3.8,9.0	269.3	22.5	73.2	2	A1
STT 101Ap	0931.9	0942N	5.4,8.4	76.3	36.6	73.0	2	A1
STF 1374	0941.4	3856N	7.3,8.6	295.9	3.22	42.68	5	D1
A.C.5	0952.5	0806S	5.8,6.2	35.	0.53	42.28	2	D1
A.C.5	0952.5	0806S	5.8,6.2	97.	0.5	73.30	1	D4
STF 1399	0957.1	1946N	7.7,9.6	175.2	30.7	73.1	2	A1
STT 215	1016.3	1744N	7.3,7.5	187.1	1.23	67.14	5	D3
				184.7	1.07	71.08	3	D4
				188.8	1.37	73.31	2	D4
STF 1415	1017.9	7104N	6.7,7.3	166.7	16.8	73.2	2	A1
STF 1424	1020.0	1951N	2.6,3.8	120.9	4.33	65.81	3,4	D3
				123.7	4.42	70.40	2	A2
				124.8	4.29	72.34	3	D4
STF 1426	1020.5	0626N	8.4,8.9	296.2	0.97	43.30	3	D1
STF 1428	1026.0	5238N	7.9,8.4	86.8	3.16	67.31	5	D3

PAIR	2000 RA	DEC	MAGS	PA	SEP	DATE	N	OBS
HU 879	1027.9	3643N	4.5,7.0	233.2	0.48	42.66	6	D1
STF 1439	1029.4	2049N	8.9,9.4	282.6	1.77	43.29	2	D1
STF 1450	1035.1	0839N	5.8,8.5	156.1	2.41	73.2	3	A1
STF 1457	1038.7	0544N	8.0,9.0	323.3	1.65	42.33	2	D1
STT 228	1047.4	2235N	8.6,9.5	175.7	0.59	67.29	4	D3
STT 229	1048.1	4107N	7.4,7.8	287.6	0.75	65.93	1	D3
				289.3	0.89	67.29	3	D3
STF 1487	1055.6	2445N	4.5,6.3	109.7	6.51	43.85	2	D1
				113.2	6.80	73.37	1	D4
BU 1077	1103.7	6145N	2.0,4.8	290.	0.73	42.40	6	D1
				284.0	0.73	43.36	9	D1
HO 378	1105.0	3825N	8.3,8.5	237.4	0.64	67.30	3	D3
STF 1517	1113.7	2008N	7.7,7.7	not seen		72.39	1	D4
				332.8	0.40	73.32	2	D4
STT 232	1115.1	3734N	8.6,9.1	241.2	0.57	67.30	2	D3
STF 1516	1115.5	7328N	7.7,8.2	103.2	49.6	73.2	3	A1
STF 1523	1118.2	3133N	4.4,4.9	256.4	1.82	43.34	12	D1
				251.0	1.78	44.38	8	D1
				245.6	1.76	45.29	3	D1
				144.0	2.42	63.30	5,6	D2
				137.3	2.55	64.01	13,14	D2
				137.0	2.59	64.85	5,6	D2
				136.7	2.60	65.36	7	D3
				134.0	2.64	66.05	10	D3
				131.4	2.71	66.95	5,8	D3
STF 1523	1118.2	3133N	4.4,4.9	132.6	2.76	68.25	8,6	D3
				123.7	2.91	70.35	1	A2
				120.1	3.09	72.33	4	D4
				118.7	3.02	73.23	3	D4
				116.6	3.10	74.35	3	D4
STF 1524	1118.5	3305N	3.7,10.1	147.3	7.51	73.2	3	A1
STF 1527	1119.1	1416N	6.9,8.1	23.2	2.50	43.39	3	D1
				32.1	2.00	73.11	3	A1
STF 1536	1123.8	1032N	4.1,7.3	299.7	0.66	43.36	9	D1
				297.9	0.62	44.39	3	D1
STF 1543	1129.1	3920N	5.4,8.4	358.9	5.51	43.35	1	D1
				358.6	5.33	73.2	3	A1
STT 234	1130.8	4117N	7.6,8.0	225.	0.3	43.39	3	D1
STF 1547	1131.8	1422N	6.4,8.4	329.2	15.2	73.2	3	A1
STT 235	1132.4	6105N	5.8,7.1	39.0	0.84	43.35	5	D1
				88.0	0.79	65.57	5	D3
				98.	0.89	67.02	3	D3
STF 1552	1134.7	1648N	6.1,7.4	207.3	3.48	67.32	5	D3
STF 1555	1136.3	2747N	6.4,6.8	131.1	0.5e	64.0	4,1	D2
				142.	0.43	65.67	4	D3
				139.0	0.48	67.06	8	D3
				149.	0.5	70.94	1	D4

PAIR	RA 2000	DEC	MAGS	PA	SEP	DATE	N	OBS
STF 1555	1136.3	2747N	6.4,6.8	143.0	0.61	73.32	4	D4
STF 1559	1138.8	6421N	6.8,7.8	323.1	1.67	73.37	2	D4
STF 1561	1138.9	4507N	6.4,8.5	255.8	9.76	43.37	2	D1
				252.3	9.64	73.2	3	A1
STT 237	1139.0	4109N	8.4,10.0	256.7	1.59	42.85	2	D1
BU 794	1153.7	7346N	7.1,8.4	73.8	0.50	43.42	2	D1
				126.0	0.47	66.0	2	D3
A 1088	1200.5	6912N	7.8,8.5	round		66.0	1	D3
STF 1596	1204.3	2128N	6.0,7.5	238.5	3.51	43.37	1	D1
STF 3123	1206.1	6842N	7.9,7.9	129.9	0.41	43.41	3	D1
STF 1604AB	1208.4	1151S	6.9,9.4	89.0	9.15	73.1	2	A1
AC	"	"	,9.3	66.6	15.55	73.1	2	A1
STF 1606	1210.8	3954N	7.3,8.0	311.7	0.76	42.40	4	D1
				313.8	0.77	43.39	4	D1
				292.0	0.62	66.0	4	D3
				289.3	0.41	70.41	1	A2
				280.4	0.66	73.36	2	D4
STF 1622	1216.1	4040N	5.9,8.2	260.6	11.6	73.2	2	A1
STF 1627	1218.1	0356S	6.6,6.9	196.2	20.2	73.1	2	A1
STF 1633	1220.6	2704N	7.0,7.1	245.1	9.01	73.1	2	A1
STT 249	1223.8	5410N	8.1,8.9	289.7	0.42	43.38	2	D1
				278.0	0.31	67.33	3	D3
				268.5	0.4	73.33	1	D4
STF 1639	1224.4	2535N	6.6,7.8	329.2	1.31	67.04	5	D3
STF 1639	1224.4	2535N	6.6,7.8	327.0	1.32	73.32	2	D4
STT 250	1224.4	4306N	8.4,8.7	346.4	0.43	43.38	1	D1
				340.	0.34	67.33	4	D3
STF 1643	1227.2	2702N	9.2,9.5	195.6	2.64	64.4	3	D2
STF 1647	1230.6	0943N	8.5,8.8	232.5	1.41	43.36	2	D1
				238.8	1.31	67.12	4	D3
STF 1657	1235.1	1823N	5.2,6.7	270.7	20.3	73.2	2	A1
STF 1661	1236.1	1124N	9.1,9.1	245.4	2.52	43.38	1	D1
STF 1670	1241.7	0127S	3.6,3.7	307.5	4.97	63.11	7,6	D2
				303.	4.87	64.93	1,2	D2
				305.8	4.65	65.4	3,1	D3
				304.9	4.74	65.9	3,4	D3
STF 1670	1241.7	0127S	3.6,3.7	305.1	4.73	66.98	4,6	D3
				303.5	4.59	68.36	1,2	D3
				304.3	4.43	70.40	1	A2
				301.4	4.32	73.41	3	D4
				301.2	4.52	74.40	3,1	D4
STF 1678	1245.4	1422N	6.8,8.5	178.0	34.8	73.1	2	A1
STF 1687	1253.3	2115N	5.2,8.0	120.2	0.94	42.68	7	D1
				117.3	1.00	44.42	1	D1
STF 1689	1255.5	1130N	7.1,9.4	217.4	29.2	73.1	3	A1
STF 1692	1256.1	3818N	2.9,5.4	228.4	19.3	73.1	2	A1
STF 1695	1256.3	5405N	6.0,7.9	279.9	3.55	67.40	2	D3

PAIR	2000 RA	DEC	MAGS	PA	SEP	DATE	N	OBS
STT 256	1256.4	0057S	7.2,7.6	89.2	0.73	42.39	3	D1
				94.1	0.72	74.37	3,2	D4
STF 1699	1258.7	2728N	8.6,8.6	7.3	1.62	67.31	4	D3
BU 1082	1300.6	5622N	4.9,8.5	317.2	0.93	43.38	7	D1
BU 929	1303.9	0340S	7.1,7.4	206.9	0.57	43.39	3	D1
				198.4	0.65	73.41	4	D4
STF 1728	1310.0	1731N	5.2,5.2	194.6	0.45	43.45	3	D1
				197.	0.44	44.43	4	D1
				15.3	0.47	65.35	6	D3
				11.3	0.60	66.0	6,4	D3
				17.	0.64	67.02	2	D3
				189.8	0.36	70.38	1	A2
STF 1728	1310.0	1731N	5.2,5.2	15.0	0.25	71.04	4	D4
STT 261	1312.0	3205N	7.2,7.7	340.7	2.19	67.09	3	D3
				341.9	2.19	70.38	1	A2
STF 1734	1320.7	0257N	6.7,7.4	180.0	1.12	67.42	1	D3
				181.5	1.07	70.40	1	A2
				180.6	1.18	73.40	2	D4
STF 1744	1323.9	5456N	2.4,4.0	151.2	14.4	73.2	2	A1
STF 1742	1324.3	0124N	7.6,8.1	0.	1.38	73.40	2	D4
STT 269	1332.8	3454N	7.3,7.8	221.8	0.33	65.24	3,2	D3
				232.2	0.36	67.24	2	D3
				236.	0.3	70.98	2	D4
				235.	0.3	73.36	1	D4
STT 269	1332.8	3454N	7.3,7.8	231.7	0.3	74.40	1	D4
STF 1757	1334.3	0009S	7.7,8.8	108.0	2.45	67.42	4	D3
STF 1768	1337.5	3617N	5.1,7.0	109.0	1.69	65.95	4	D3
				110.7	1.64	67.26	4	D3
				103.2	1.52	70.43	1	A2
				110.4	1.70	72.46	3	D4
STF 1770	1337.8	5042N	6.8,8.3	121.3	1.61	67.22	5	D3
				119.0	1.31	70.42	1	A2
BU 612	1339.6	1044N	6.3,6.3	218.	0.38	65.98	1	D3
				227.1	0.32	67.30	4	D3
				240.	0.25	71.08	1	D4
STF 1781	1346.1	0507N	7.8,8.2	7.0	0.55	67.31	3	D3
STF 1785	1349.2	2659N	7.9,8.2	118.0	1.99	42.42	7	D1
				119.7	2.05	43.46	7	D1
				124.2	2.13	44.54	5	D1
				146.6	3.12	63.14	6,5	D2
				147.4	3.14	66.1	3	D3
				147.8	3.14	67.15	5	D3
				151.9	3.30	70.42	1	A2
				155.0	3.27	72.41	3	D4
				153.6	3.27	73.21	3	D4
				155.8	3.31	74.28	3	D4
STT 276	1408.2	3644N	8.6,9.4	203.7	0.39	67.36	3	D3

PAIR	2000 RA	DEC	MAGS	PA	SEP	DATE	N	OBS
STT 278	1412.0	4411N	8.4,8.6	333.8	0.32	67.28	3	D3
STT 277	1412.4	2843N	8.3,8.5	41.7	0.41	73.44	2	D4
STF 1820	1413.1	5520N	8.8,9.1	95.8	2.39	43.54	1	D1
				102.1	2.32	74.40	4,3	D4
STF 1821	1413.5	5147N	4.6,6.6	236.3	13.5	73.2	1	A1
STF 1816	1413.9	2906N	7.5,7.6	84.3	0.94	66.1	6,3	D3
				91.0	0.83	67.10	3	D3
				86.1	0.86	70.40	1	A2
STF 1817	1414.2	2642N	8.9,9.5	1.5	0.50	67.14	6	D3
STF 1834	1420.3	4831N	7.9,8.0	97.0	0.68	44.07	4	D1
				102.2	1.09	67.09	4	D3
A 570	1432.3	2641N	6.6,6.8	318.5	0.25	70.41	1	A2
STF 1863	1438.1	5135N	7.4,7.7	77.8	0.58	44.07	4	D1
				69.2	0.65	74.35	2	D4
STF 1864	1440.7	1626N	4.9,5.8	107.5	5.77	43.02	6	D1
				104.8	5.77	66.0	2	D3
				108.9	5.58	67.01	6,8	D3
STF 1865	1441.1	1344N	4.4,4.8	132.2	1.08	44.57	5	D1
				129.5	1.35	70.48	5	D4
				126.8	1.07	70.41	3	A2
				126.5	1.27	73.44	2	D4
STF 1871	1441.4	5124N	8.0,8.0	304.5	1.86	67.28	3	D3
				304.3	1.84	73.42	2	D4
STF 1877	1445.0	2704N	2.7,5.1	338.5	2.78	65.6	3,6	D3
STF 1877	1445.0	2704N	2.7,5.1	337.0	2.77	67.09	2	D3
				332.4	2.70	73.43	4	D4
STT 285	1445.5	4222N	7.7,8.2	55.4	0.41	43.49	5	D1
				57.2	0.49	44.54	3	D1
				326.	0.3	67.16	5	D3
STF 1884	1448.5	2422N	6.2,7.8	58.3	1.93	70.42	1	A2
BU 106	1449.3	1409S	5.8,6.7	0.3	1.71	73.43	3	D4
STF 1890	1449.6	4843N	6.1,6.8	44.1	2.90	66.1	6,3	D3
				42.4	2.95	67.45	5	D3
STT 287	1451.4	4455N	8.5,8.6	157.4	1.03	42.68	5	D1
				164.2	1.12	67.16	3	D3
STF 1888	1451.4	1907N	4.8,6.7	4.7	5.61	43.47	11	D1
STF 1888	1451.4	1907N	4.8,6.7	3.2	5.68	44.52	7	D1
				347.0	6.87	64.40	2	D2
				347.2	7.06	64.5	1,2	D2
				342.4	7.00	66.1	3,4	D3
				342.2	6.92	67.22	4	D3
				341.8	7.04	70.38	1	A2
				339.2	7.30	70.51	3	D4
				336.2	7.10	73.33	3	D4
				336.3	7.36	74.42	2	D4
STT 288	1453.4	1543N	6.9,7.6	176.0	1.55	67.16	3	D3
				178.4	1.63	70.37	1	A2

PAIR	2000 RA	DEC	MAGS	PA	SEP	DATE	N	OBS
BU 348	1501.8	0008S	6.0,8.3	107.3	0.39	70.51	1	A2
STF 1909	1503.9	4739N	5.3,6.2	251.8	2.13	43.45	16	D1
				252.4	2.05	44.48	12	D1
				253.5	2.00	45.74	5	D1
				279.5	0.83	63.30	9	D2
				283.0	0.64	65.3	13,12	D3
				291.8	0.64	66.1	5,4	D3
				301.0	0.56	67.22	5	D3
				313.1	0.53	68.49	5	D3
				334.2	0.50	70.48	4	D4
				335.9	0.47	70.44	3	A2
				347.	0.5	71.52	1	D4
STF 1909	1503.9	4739N	5.3,6.2	356.4	0.48	72.43	5	D4
				1.5	0.60	73.35	6	D4
				6.8	0.53	74.36	3	D4
A 1116	1511.6	1008N	8.5,8.5	40.4	0.47	42.52	1	D1
STF 1932	1518.3	2649N	7.1,7.6	240.4	1.04	64.4	2	D2
				241.6	1.14	65.8	1	D3
				243.5	1.07	67.16	3	D3
				244.5	1.0	71.06	2	D4
				247.1	1.16	73.50	3	D4
BU 32	1521.1	0044N	5.5,10.1	22.3	2.98	70.41	1	A2
STF 1937	1523.2	3018N	5.6,6.1	11.3	0.87	44.50	7	D1
				119.6	0.53	64.38	2	D2
STF 1937	1523.2	3018N	5.6,6.1	124.2	0.6e	64.52	2	D2
				132.2	0.54	65.27	10	D3
				144.8	0.55	66.15	5	D3
				152.9	0.57	67.17	5	D3
				165.7	0.59	68.40	6	D3
				183.2	0.62	70.39	2	A2
				183.3	0.63	70.48	5	D4
				190.5	0.65	71.45	3	D4
				200.0	0.59	72.38	4	D4
				208.6	0.62	73.34	6	D4
				213.2	0.68	74.25	3	D4
STF 1938	1524.5	3721N	7.2,7.8	23.6	2.03	63.04	6,4	D2
				20.6	2.08	70.38	2	A2
				19.4	2.07	72.50	4	D4
				20.6	2.01	73.42	4	D4
HU 149	1524.6	5413N	7.5,7.6	276.1	0.44	70.51	1	A2
STT 296	1526.5	4400N	7.6,9.2	283.5	1.69	70.50	1	A2
STF 1956	1533.2	4149N	8.5,10.0	not seen		73.46	5	D4
STF 1954	1534.8	1032N	4.2,5.2	181.0	3.76	43.0	12	D1
				179.2	3.95	67.47	5	D3
				183.1	4.04	70.42	1	A2
STT 298	1536.1	3948N	7.4,7.7	121.	0.3	42.50	4	D1
				146.7	0.36	43.48	4	D1

PAIR	2000 RA	DEC	MAGS	PA	SEP	DATE	N	OBS
STT 298	1536.1	3948N	7.4,7.7	156.6	0.58	44.54	5	D1
				162.5	0.55	45.73	3	D1
				189.2	1.28	66.1	3	D3
				192.9	1.13	67.45	4	D3
				195.9	0.91	70.41	3	A2
				194.6	1.12	70.51	4	D4
				205.3	1.04	73.42	6	D4
STF 1963	1537.9	3006N	8.8,9.2	296.3	5.14	67.12	3	D3
STF 1965	1539.4	3638N	5.1,6.0	306.0	6.20	67.59	5,8	D3
STF 1989	1539.7	7958N	7.3,8.3	30.0	0.65	67.28	5	D3
STF 1969	1541.4	5959N	8.9,9.6	round		66.2	1	D3
HU 580	1541.6	1941N	5.3,5.3	257.0	0.20	70.41	1	A2
STF 1967	1542.8	2618N	4.0,7.0	291.0	0.60	43.57	4	D1
BU 619	1543.1	1340N	6.9,7.4	1.6	0.60	67.28	3	D3
				3.6	0.53	70.41	1	A2
				8.5	0.57	73.52	3	D4
STT 303	1600.9	1315N	7.5,8.0	167.2	1.37	73.52	3	D4
STF 1998AB	1604.4	1122S	4.8,5.1	40.8	0.70	44.05	1	D1
				0.3	1.18	67.53	3	D3
				7.5	1.34	72.52	2	D4
				11.0	1.17	73.49	3	D4
AC	"	"	4.8,7.2	44.7	7.50	73.54	2	D4
BC	"	"	5.1,7.2	51.0	6.75	73.54	2	D4
STF 1999	1604.4	1126S	7.4,8.1	100.4	10.92	73.53	2	D4
BU 812	1607.1	1654N	9.2,9.3	112.5	0.70	43.54	1	D1
BU 120 AB	1612.1	1927S	4.4,6.9	357.8	1.03	73.51	3	D4
CD	"	"	6.9,7.9	49.0	1.4	73.51	3	D4
STF 2021	1613.3	1333N	7.5,7.7	348.8	4.88	70.43	1	A2
STF 2032	1614.7	3352N	5.8,6.7	225.6	5.75	43.04	13	D1
				229.6	6.32	63.3	7,4	D2
				229.0	6.24	66.1	4,1	D3
				230.7	6.22	67.95	10	D3
				230.8	6.35	70.50	3	A2
STT 309	1619.2	4140N	8.6,8.8	284.	0.33	67.28	4	D3
STF 2054	1623.8	6141N	5.9,7.1	354.1	0.98	67.53	3	D3
				355.6	1.02	70.41	1	A2
				351.7	1.12	73.51	3	D4
STF 2049	1628.0	2559N	7.1,8.1	202.4	0.97	73.40	2	D4
STF 2052	1628.9	1825N	7.8,7.8	209.2	0.62	43.55	5	D1
				162.8	0.93	63.15	2,3	D2
				147.2	1.06	67.20	2	D3
				147.8	0.93	70.46	3	A2
				140.5	1.20	72.52	3	D4
				141.6	1.0	73.47	3	D4
STF 2055	1631.0	0159N	3.9,6.0	239.1	0.43	42.59	4	D1
				176.6	0.99	67.15	2	D3
				4.1	0.87	70.38	1	A2

PAIR	2000 RA	DEC	MAGS	PA	SEP	DATE	N	OBS
STF 2055	1631.0	0159N	3.9,6.0	5.2	1.23	73.34	5	D4
STT 313	1632.6	4006N	7.7,8.3	137.3	0.88	65.3	5,4	D3
				135.6	0.92	67.44	3	D3
				136.9	0.72	73.44	3	D4
STF 2078	1636.3	5255N	5.6,6.6	106.6	3.22	65.3	1	D3
				107.9	3.33	67.98	4,5	D3
STF 2084	1641.4	3136N	3.0,6.5	33.7	0.99	64.4	1	D2
				20.3	0.70	65.3	2	D3
				5.5	0.65	65.60	5,4	D3
				356.	0.56	66.2	1	D3
				309.2	0.48	67.22	6,2	D3
				206.7	1.2	72.47	6,1	D4
STF 2084	1641.4	3136N	3.0,6.5	195.3	1.15	73.52	5	D4
STF 2091	1642.1	4112N	8.3,8.8	311.8	0.65	67.56	2	D3
HU 664	1643.7	5132N	8.4,8.4	306.2	0.38	72.56	3	D4
D 15	1643.9	4329N	9.1,9.1	152.3	1.21	67.56	2	D3
				144.0	1.17	73.57	2	D4
STF 2094	1644.2	2331N	7.4,7.7	72.7	1.23	67.16	2	D3
STF 2106	1651.1	0925N	7.0,8.7	189.7	0.53	73.56	2	D4
STT 315	1651.5	0113N	5.7,7.6	not seen		73.55	4	D4
STF 2107	1651.9	2840N	6.7,8.2	79.6	1.20	65.9	4	D3
				83.7	1.08	70.57	2	A2
				85.4	1.44	73.45	2	D4
STF 2118	1656.3	6502N	6.9,7.4	74.0	1.02	65.6	2	D3
STF 2118	1656.3	6502N	6.9,7.4	71.6	1.07	67.39	6	D3
				71.5	1.12	73.56	2	D4
STF 2114	1701.9	0827N	6.5,7.7	183.0	1.16	70.54	2	A2
				187.	1.5	73.56	3	D4
STF 2130	1705.4	5427N	5.8,5.8	96.0	2.20	43.55	12	D1
				95.1	2.16	44.61	5	D1
				69.9	2.15	63.1	2	D2
				67.6	2.01	65.3	3	D3
				67.5	2.01	65.6	3	D3
				63.4	2.10	67.20	5	D3
				63.9	2.14	68.43	5	D3
				55.6	2.10	72.32	4	D4
STF 2130	1705.4	5427N	5.8,5.8	56.0	2.15	73.57	4,1	D4
KUI 79	1712.1	4544N	10.1,10.6	272.8	0.76	67.34	4	D3
				231.2	1.0	73.55	3	D4
STF 2140	1714.6	1423N	3.5,5.4	111.7	4.66	43.63	2	D1
				106.9	4.61	70.6	3	A2
STF 2161	1723.7	3708N	4.5,5.5	316.0	4.00	67.68	4,6	D3
BU 1201	1726.3	6746N	8.8,8.8	356.	0.35	72.57	2	D4
HO 417	1729.3	3758N	9.3,9.3	147.7	0.37	42.66	4	D1
STF 2173	1730.3	0103S	5.9,6.2	334.5	1.03	42.64	14	D1
				164.2	0.82	63.12	3,2	D2
				131.4	0.32	73.39	5	D4

PAIR	2000 RA	DEC	MAGS	PA	SEP	DATE	N	OBS
STF 2207	1737.0	6707N	8.3,8.8	119.1	0.44	67.30	2	D3
				105.5	0.52	73.60	3	D4
STF 2199	1738.6	5546N	7.8,8.4	75.9	1.70	42.15	9	D1
				70.5	1.83	72.56	3	D4
STF 2218	1740.3	6340N	7.1,8.3	326.7	1.53	65.6	2	D3
				326.9	1.63	67.28	6	D3
				327.4	1.86	72.56	3	D4
STF 2203	1741.2	4140N	7.6,6.9	302.9	0.85	67.35	3	D3
				301.4	0.86	72.56	2	D4
A.C.7	1746.5	2745N	10.2,10.7	53.6	1.39	42.62	3	D1
				250.0	1.13	67.36	4,1	D3
				310.	0.56	73.52	5	D4
STF 2215	1747.2	1742N	5.8,7.8	272.9	0.64	67.24	3	D3
STT 337	1750.5	0715N	8.2,8.7	round		43.65	1	D1
				181.9	0.27	73.59	2	D4
STF 2242	1751.2	4455N	8.0,8.0	326.8	3.48	67.33	3	D3
STT 338	1752.0	1520N	6.8,7.1	5.2	0.80	43.65	2	D1
				358.	0.84	65.33	1	D3
				355.6	0.83	67.59	2	D3
				357.	1.10	73.34	2	D4
STF 2264	1801.5	2136N	5.1,5.2	258.	6.33	68.58	2,4	D3
STF 2267	1801.6	4011N	8.6,8.6	254.0	0.73	67.29	3	D3
STF 2262	1803.0	0811S	5.3,6.0	273.7	1.93	72.60	2	D4
				269.5	1.92	73.53	3	D4
STF 2272	1805.4	0232N	4.3,6.3	110.7	6.60	43.68	6	D1
				110.7	6.37	44.60	2	D1
				87.8	3.84	63.12	9,10	D2
				82.	3.76	64.38	1	D2
				80.9	3.56	64.53	2,4	D2
				78.3	3.28	65.57	5,6	D3
				78.0	3.13	66.1	3	D3
				72.0	2.87	67.34	6	D3
				70.0	2.65	68.31	10	D3
				40.0	2.03	72.24	4	D4
				24.9	1.96	73.23	4	D4
STT 341	1805.9	2126N	7.2,8.5	91.2	0.35	43.70	2	D1
				87.	0.41	66.5	3,2	D3
				96.	0.39	67.38	3,2	D3
				91.3	0.38	72.59	3	D4
				95.1	0.37	73.60	3	D4
HU 1186	1806.3	3824N	8.7,8.8	96.1	0.39	72.59	2	D4
STF 2281	1809.5	0401N	5.9,7.4	358.9	0.42	67.37	4	D3
				349.6	0.32	73.60	3	D4
HU 674	1809.7	5024N	7.5,8.0	232.7	0.72	72.60	2	D4
STF 2289	1810.2	1628N	6.4,7.5	224.2	1.15	67.59	3	D3
				224.8	1.14	70.58	3	A2
				222.9	1.10	73.60	3	D4

PAIR	2000 RA	DEC	MAGS	PA	SEP	DATE	N	OBS
STF 2294	1814.5	0010N	8.5,8.8	98.3	0.94	67.37	3	D3
				97.3	0.83	73.62	3	D4
STT 353	1820.7	7120N	4.4,6.1	round		65.6	1	D3
A.C.11	1824.9	0135S	6.8,7.0	358.6	0.75	67.63	2	D3
				0.7	1.0	73.60	2	D4
STF 2315	1825.0	2723N	6.6,7.6	150.4	0.42	43.67	3	D1
				136.7	0.47	66.1	5	D3
				136.3	0.54	67.41	4	D3
				133.2	0.63	72.61	3	D4
				137.1	0.52	73.50	3	D4
HU 66	1825.3	4845N	7.9,8.1	274.0	0.38	43.78	1	D1
				265.0	0.25	73.53	2	D4
STT 351	1825.3	4845N	7.7,8.2	22.0	0.7	43.78	1	D1
				22.0	0.70	73.53	2	D4
STF 2339	1833.8	1744N	7.1,8.0	272.7	1.98	70.57	3	A2
A 1377 AB	1834.0	5221N	6.2,6.2	82.1	0.29	73.63	3	D4
AB - C	"	"	5.4,8.8	269.6	26.01	73.63	3	D4
STT 359	1835.5	2336N	6.4,6.7	185.7	0.53	65.8	2	D3
				195.1	0.51	67.50	4	D3
				193.7	0.56	70.57	3	A2
				197.0	0.60	72.60	1	D4
				191.7	0.53	73.45	3	D4
STT 358	1835.9	1659N	6.8,7.2	167.6	1.73	70.57	3	A2
				168.0	1.47	73.63	1	D4
STT 357	1835.9	1144N	8.1,8.2	104.	0.27	73.59	2	D4
STF 2351	1836.2	4115N	7.7,7.7	340.9	5.22	43.63	8	D1
STF 2384	1838.6	6708N	8.6,9.1	307.8	0.71	67.62	3	D3
STF 2368	1838.9	5221N	7.6,7.8	323.6	1.85	73.46	2	D4
HO 437	1840.6	3138N	8.4,8.6	119.	0.45	43.78	1	D1
STF 2367	1841.3	3018N	7.4,7.9	56.	0.36	67.52	4	D3
				round		73.36	1	D4
STF 2398	1843.3	5933N	9.3,9.8	160.8	16.40	45.80	1	D1
				166.7	14.84	73.68	3	D4
STF 2382	1844.3	3940N	5.1,6.1	0.4	2.69	65.7	4	D3
				358.7	2.71	67.38	4	D3
				358.5	2.64	70.48	3	A2
STF 2382	1844.3	3940N	5.1,6.1	355.7	2.60	73.60	4	D4
STF 2383	1844.3	3940N	5.1,5.4	98.5	2.28	65.7	2,5	D3
				96.0	2.31	67.38	4	D3
				97.6	2.31	70.48	3	A2
				91.5	2.36	73.60	4	D4
STF 2415	1854.5	2037N	7.0,8.9	292.9	1.63	73.59	3	D4
BU 648	1857.1	3253N	5.3,7.5	288.5	0.79	42.65	5	D1
				287.8	0.75	43.74	2	D1
				274.9	0.73	44.60	4	D1
				273.3	1.02	45.70	3	D1
STF 2422	1857.2	2605N	8.0,8.1	80.7	0.80	73.42	4	D4

PAIR	2000 RA	DEC	MAGS	PA	SEP	DATE	N	OBS
STF 2438	1857.5	5814N	6.8,7.4	3.5	0.87	73.43	3	D4
STF 2437	1901.8	1911N	8.2,8.4	42.0	0.65	43.76	2	D1
				not seen		73.59	3	D4
SHJ 286	1904.9	0402S	5.5,7.2	209.6	39.07	73.71	3	D4
A 863	1909.8	0018S	9.4,10.7	round		66.0	1	D3
STF 2481	1911.1	3847N	8.3,8.3	210.5	4.40	73.65	1	D4
SE 2	"	"	8.3,9.3	69.	0.35	42.76	2	D1
				round		66.0	1	D3
				102.	0.38	73.65	2	D4
STF 2486	1912.1	4950N	6.6,6.8	213.5	8.57	45.18	3	D1
				211.0	8.17	66.4	2,1	D3
				209.7	8.05	67.78	6	D3
STF 2486	1912.1	4950N	6.6,6.8	208.1	8.29	73.70	2	D4
BU 139	1912.6	1651N	6.7,8.0	313.3	0.70	73.62	3	D4
STT 371	1915.9	2727N	7.0,7.1	338.0	0.93	65.9	6	D3
				156.2	0.67	73.61	2	D4
STF 2491	1916.2	2816N	8.4,9.7	50.	1.20	73.60	3,1	D4
STF 2509	1916.9	6313N	7.2,8.3	328.5	1.54	67.63	3	D3
BU 1129	1921.6	5223N	7.7,7.7	round		65.79	1	D3
STF 2525	1926.5	2719N	8.5,8.7	295.1	1.60	67.59	4	D3
STT 377	1936.2	3539N	9.3,9.4	215.9	1.0	43.78	1	D1
STF 2556	1939.4	2216N	7.7,8.2	89.4	0.32	43.67	6	D1
				51.	0.36	65.8	5	D3
STF 2574	1940.5	6240N	8.1,8.1	248.1	0.34	73.69	3	D4
STT 380	1942.6	1149N	5.6,6.8	82.5	0.51	44.76	4	D1
STT 383	1943.0	4043N	6.9,8.4	20.8	0.82	43.44	4	D1
STT 384	1943.7	3819N	7.6,7.9	194.9	1.12	65.8	1	D3
STF 2579	1945.0	4507N	3.0,7.9	246.7	2.20	65.78	3,4	D3
				240.3	2.19	70.62	3	A2
STF 2576	1945.6	3336N	9.3,9.3	178.8	0.48	43.70	7	D1
				159.4	0.47	44.62	5	D1
				133.5	0.47	45.74	4	D1
				12.4	1.48	67.74	3	D3
				6.6	1.67	73.43	3	D4
STT 386	1948.2	3710N	8.2,8.5	74.1	0.94	67.27	3	D3
STF 2603	1948.2	7016N	4.0,7.6	10.9	3.02	44.11	3	D1
STF 2603	1948.2	7016N	4.0,7.6	10.4	3.08	65.82	3,2	D3
STF 2583	1948.7	1148N	6.1,6.9	108.3	1.43	70.64	3	A2
STT 387	1948.7	3519N	6.9,7.9	257.5	0.43	43.21	10	D1
				188.5	0.6	65.8	2	D3
				189.2	0.53	67.06	4	D3
A.G.C.11	1948.9	1908N	5.4,6.4	147.8	0.20	70.63	1	A2
HO 581	1954.9	4152N	7.9,8.4	231.2	0.3	42.71	2	D1
STF 2605	1955.6	5226N	4.9,7.4	179.9	3.11	67.44	4	D3
STF 2606	1958.6	3316N	7.6,8.3	139.4	0.92	67.73	4	D3
STT 395	2001.8	2456N	5.9,6.3	111.4	0.77	42.70	4	D1
				120.1	0.86	64.8	2	D2

PAIR	2000 RA	DEC	MAGS	PA	SEP	DATE	N	OBS
STT 395	2001.8	2456N	5.9,6.3	117.4	0.87	65.8	3	D3
				117.0	0.85	67.27	4,3	D3
				116.0	0.75	70.60	3	A2
				118.7	0.87	73.43	2	D4
STF 2624	2003.5	3602N	7.2,7.8	175.2	1.90	43.77	4	D1
				174.0	2.07	67.20	3	D3
STF 2626	2004.2	3031N	8.9,9.1	125.5	0.98	67.35	4	D3
STF 2642	2005.5	6342N	9.6,9.6	359.8	2.20	44.05	4	D1
STF 2652	2009.0	6205N	7.2,7.5	226.5	0.33	66.80	5,3	D3
STT 400	2010.2	4357N	7.5,8.5	319.1	0.62	43.76	4	D1
				not seen		73.5	1	D4
STT 403	2014.3	4206N	7.1,7.3	173.1	0.82	44.28	6	D1
STT 403	2014.3	4206N	7.1,7.3	173.	0.6	73.42	1	D4
STF 2671	2018.4	5524N	6.0,7.4	336.2	3.50	65.8	2	D3
				338.0	3.38	67.72	3	D3
STT 406	2019.9	4522N	7.4,8.3	123.0	0.66	65.72	1	D3
				122.	0.55	67.50	4	D3
				113.0	0.53	73.66	4	D4
D 22	2025.5	4005N	8.0,9.1	151.1	2.92	42.78	1	D1
HO 130	2026.2	3712N	9.2,9.4	286.7	1.57	67.53	4	D3
BU 151	2037.5	1416N	4.1,5.1	331.	0.55	44.72	5	D1
				336.6	0.55	45.75	5	D1
				318.3	0.39	70.72	1	A2
				343.7	0.56	73.40	2	D4
STF 2705	2037.7	3322N	7.4,8.4	264.7	3.02	43.77	3	D1
STT 533	2039.1	1005N	5.2,11.8	282.2	34.7	70.72	1	A2
STT 410	2039.6	4036N	6.5,6.8	185.0	0.84	64.8	2	D2
				6.3	0.84	65.8	2	D3
				9.1	0.81	67.61	3	D3
				9.7	0.70	70.63	3	A2
STF 2716	2041.1	3219N	5.9,8.0	45.1	2.59	43.80	3	D1
STF 2723	2045.0	1219N	6.9,8.7	110.7	1.17	44.00	4	D1
STF 2726	2045.6	3043N	4.3,9.5	61.4	6.25	44.77	3	D1
BU 676	2046.2	3358N	2.6,11.6	268.9	47.3	70.72	1	A2
STF 2725	2046.3	1554N	7.5,8.2	5.4	5.42	43.74	1	D1
STF 2727	2046.7	1608N	4.5,5.5	269.	10.05	64.6	1,4	D2
STF 2727	2046.7	1608N	4.5,5.5	270.0	9.75	65.8	2,4	D3
				269.1	9.68	67.90	2	D3
STT 413	2047.4	3629N	4.8,6.1	34.1	0.74	45.82	3	D1
				29.5	0.75	64.8	2	D1
				24.1	0.77	70.70	4	A2
				26.2	0.73	73.73	3	D4
BU 268	2047.6	4204N	7.7,8.6	213.4	0.45	42.78	1	D1
BU 155	2051.1	5125N	7.3,8.0	38.6	0.79	65.76	2	D3
				37.2	0.91	67.20	4	D3
HO 144	2052.3	2008N	8.0,8.0	166.9	0.3	67.36	4	D3
HO 146	2053.6	3514N	8.7,8.7	50.2	0.28	67.55	6	D3

PAIR	2000 RA	DEC	MAGS	PA	SEP	DATE	N	OBS
STT 418	2054.8	3242N	8.1,8.2	107.4	1.18	65.78	2	D3
				286.3	1.05	67.48	3	D3
STF 2741	2058.6	5028N	5.8,7.1	27.6	2.07	42.92	3	D1
				24.3	1.85	67.22	4,6	D3
STF 2737	2059.1	0418N	5.8,6.3	284.1	0.98	65.8	1,2	D3
				287.0	1.04	70.60	3	A2
				285.0	1.00	73.43	2	D4
STF 2742	2102.2	0711N	7.4,7.4	220.9	2.67	43.72	1	D1
				216.3	2.73	67.27	3	D3
STF 2751	2102.2	5640N	6.1,7.1	171.8	1.87	65.8	3	D3
				350.9	1.60	67.25	3	D3
STF 2744	2103.1	0132N	7.0,7.5	129.5	1.38	70.67	4	A2
STF 2758	2106.3	3839N	5.6,6.3	138.6	26.69	45.77	3	D1
				145.0	28.67	73.72	4	D4
STF 2760	2106.8	3408N	8.0,8.8	239.0	2.01	43.85	2	D1
				239.7	1.86	44.71	5	D1
				240.7	1.77	45.78	5	D1
				322.8	0.65e	64.68	3,2	D2
				332.5	0.64	65.70	11	D3
				343.5	0.74	66.81	2	D3
				345.7	0.58	67.56	9	D3
				350.8	0.58	68.57	3,1	D3
				17.6	0.71	73.62	5	D4
STF 2780	2111.8	6000N	6.0,7.0	215.7	1.23	70.68	3	A2
H I 48	2113.7	6425N	7.1,7.3	250.6	0.79	67.44	2	D3
STT 432	2114.4	4109N	7.8,8.2	112.9	1.35	65.76	2	D3
				116.2	1.27	67.60	3	D3
				117.9	1.33	70.67	3	A2
STT 535	2114.5	1001N	5.3,5.4	21.5	0.34	67.13	8,4	D3
				22.	0.3	73.41	2	D4
HO 286	2119.4	3815N	6.6,6.6	round		65.8	2	D3
STT 437	2120.8	3228N	6.9,7.6	208.9	2.06	64.8	1,3	D2
				28.0	2.41	65.72	3,2	D3
				27.2	2.26	70.67	3	A2
STT 435	2121.4	0254N	8.1,8.6	229.8	0.51	67.70	3	D3
HO 157	2123.0	3202N	9.2,9.2	22.6	3.68	42.88	1	D1
HO 157	2123.0	3202N	9.2,9.2	24.2	3.41	45.81	1	D1
				24.0	3.94	65.76	2	D3
STF 2799	2128.9	1105N	7.5,7.5	272.3	1.66	70.59	3	A2
STF 2804	2133.0	2043N	7.6,8.3	347.2	3.13	67.52	2,3	D3
STT 445	2139.3	2043N	9.2,9.7	114.	0.74	67.72	2	D3
BU 1212	2139.5	0003S	7.3,7.8	275.	0.45	65.8	3,1	D3
				306.	0.35	67.68	5	D3
HO 166	2143.9	2751N	8.8,8.8	130.	0.36	65.8	5,3	D3
				124.8	0.35	66.86	3	D3
				108.	0.3	72.58	1	D4
				107.	0.37	73.63	5	D4

PAIR	2000 RA	DEC	MAGS	PA	SEP	DATE	N	OBS
STF 2822	2144.1	2845N	4.7,6.1	253.9	0.62	43.74	3	D1
				256.2	0.71	44.65	5	D1
				264.2	0.78	45.79	4	D1
				285.2	1.68	64.77	2,1	D2
				285.5	1.75	65.75	7,5	D3
				290.1	1.84	67.47	4	D3
				288.4	1.86	70.63	3	A2
				293.5	1.95	73.57	4	D4
COU 14	2150.2	1718N	5.5,7.5	214.	0.3	67.55	6,2	D3
STF 2843	2151.6	6545N	7.1,7.3	145.4	1.64	67.73	3	D3
STT 458	2156.4	5946N	7.0,8.5	348.2	0.83	67.83	3	D3
BU 275	2157.3	6117N	7.7,7.7	351.6	0.38	67.74	4	D3
STF 2863	2203.7	6437N	4.6,6.5	274.8	7.54	73.75	3	D4
STF 2881	2214.5	2935N	7.6,8.1	91.3	1.37	42.90	1	D1
HO 180	2215.7	4354N	8.2,8.2	236.3	0.66	67.56	4	D3
BU 1216	2220.2	2931N	9.3,9.6	288.0	0.61	67.55	4	D3
BU 172	2224.1	0451S	6.6,6.6	312.9	0.47	67.61	4	D3
				301.8	0.33	73.80	3	D4
KR 60 AB	2228.1	5742N	9.4,10.9	58	3.10	65.82	1	D3
				195.3	1.17	66.72	3,1	D3
				238.8	1.6	73.80	3	D4
" AC	"	"	,10.1	62.2	102.22	73.80	2	D4
" AE	"	"	,12.5	127.7	221.60	73.80	2	D4
STF 2909	2228.8	0002S	4.4,4.6	284.4	2.50	43.82	1	D1
STF 2909	2228.8	0002S	4.4,4.6	283.0	2.33	44.86	2	D1
				281.9	2.40	45.79	5	D1
				251.8	1.93	64.71	7,12	D2
				248.2	1.92	65.82	5,7	D3
				252.0	1.97	66.92	1,6	D3
				248.1	1.91	67.47	6,4	D3
				248.0	1.93	68.56	3,4	D3
				236.8	1.79	73.62	7	D4
STF 2912	2229.9	0425N	5,8,7.2	115.9	0.73	73.78	3	D4
STF 2920	2234.5	0413N	8.1,9.2	142.3	14.0	72.9	1	A1
HO 296	2240.8	1432N	6.6,6.6	round		42.78	1	D1
				71.0	0.36	67.11	6	D3
HO 296	2240.8	1432N	6.6,6.6	70.9	0.44	68.04	2	D3
				59.6	0.42	70.64	3	A2
				53.6	0.65	72.58	1	D4
				50.3	0.45	73.59	5	D4
STF 2934	2241.8	2125N	8.7,9.7	284.0	0.85	45.30	4	D1
STF 2943	2247.7	1404S	5.8,9.0	122.4	22.65	73.79	2	D4
STF 2950	2251.3	6141N	6.1,7.4	288.9	1.71	67.82	4,6	D3
				291.1	1,8	73.53	2	D4
A 632	2252.1	5744N	8.2,9.0	191.6	0.68	42.84	3	D1
COU 240	2256.4	2257N	7.4,7.8	289.6	0.65	67.35	10	D3
				287.0	0.62	72.64	2	D4

PAIR	2000 RA	DEC	MAGS	PA	SEP	DATE	N	OBS
STT 483	2259.2	1144N	6.0,7.5	257.8	0.72	42.86	3	D1
HLD 56	2259.7	4149N	9.3,9.4	108.3	0.91	43.82	2	D1
STF 2974	2305.0	3322N	8.1,8.1	164.4	2.65	42.87	8	D1
BU 385	2310.3	3228N	7.3,8.1	100	0.67	65.81	1	D3
				95.9	0.51	70.60	1	A2
STF 2993	2314.0	0856S	8.3,9.4	176.7	25.7	72.9	1	A1
AC	"	"	,11.4	123.2	95.6	72.9	1	A1
STF 3001	2318.5	6807N	5.0,7.6	212.6	2.88	65.84	5,4	D3
				215.2	3.08	67.88	3	D3
				218.8	2.76	73.61	3	D4
BU 80	2318.9	0525N	9.0,10.0	281.5	1.05	42.90	2	D1
STF 2998	2319.1	1327S	5.5,7.5	351.3	13.1	72.9	1	A1
STT 494	2320.8	2157N	8.3,9.0	83.6	3.39	41.90	2	D1
STF 3007	2322.8	2034N	6.7,9.7	90.5	6.0	72.9	1	A1
STT 496	2330.0	5833N	4.9,9.3	218.9	0.89	42.85	2	D1
BU 720	2334.0	3120N	6.0,6.0	62.9	0.5e	64.60	3	D2
				242.8	0.47	65.6	3	D3
				251.6	0.67	67.75	4	D3
				250.1	0.50	70.71	3	A2
				72.0	0.4	70.96	2	D4
				256.2	0.5	72.58	1	D4
				250.6	0.45	73.52	5	D4
				252.6	0.6	73.83	1	D4
STT 500	2337.5	4426N	6.3,7.2	345.5	0.50	43.82	1	D1
STT 500	2337.5	4426N	6.3,7.2	167.2	0.56	65.60	4	D3
				349.8	0.53	67.9	2	D3
				353.4	0.45	70.66	4	A2
BU 858	2341.3	3234N	7.4,8.9	240.0	0.62	42.93	3	D1
STT 510	2351.6	4205N	7.7,8.0	135.7	0.49	65.6	2	D3
STF 3050	2359.5	3343N	6.6,6.6	289.5	1.53	65.82	1	D1
				114.0	1.49	70.64	3	A2
				300.5	1.63	73.60	4	D4

The following measures were received after the main catalogue had been prepared.

PAIR	RA	DEC	MAGS	PA	SEP	DATE	N	OBS
STF 98	0112.9	3205N	7.0,8.0	38.7	0.66	75.15	3	D4
STF 202	0202.0	0246N	4.3,5.2	286.5	1.91	75.09	3	D4
STF 728	0530.7	0556N	4.5,6.0	50.9	0.86	75.15	3	D4
STF 1110	0734.6	3153N	2.0,2.9	120.2	2.01	75.15	3	D4
STF 1196	0812.2	1740N	5.6,6.0	308.2	0.92	75.15	2	D4
STF 1356	0928.5	0904N	5.9,6.7	1.8	0.5	75.32	2	D4
STT 215	1016.3	1744N	7.3,7.5	187.5	1.2	75.33	2	D4
STF 1523	1118.2	3133N	4.4,4.9	116.9	3.02	75.31	3	D4
STF 2272	1805.4	0232N	4.3,6.3	12.4	1.90	75.27	3	D4

CATALOGUE OF DOUBLE STARS
(ii) Notes

PAIR	NOTES
STT 547	ADS 48. P=362 years (1975.0 = 5".88) o,o.
STF 3062	ADS 61. P=106.8 years (1975.0 = 1".41) Py,Py.
BU 255	ADS 147. Very slowly widening. 1875, 99°,0".4.
BU 1026	ADS 148. P=68.5 years (1975.0 = 0".23)
STF 12	ADS 191. 35 Psc. Fixed. y,Pb.
STT 4	ADS 221. P=112.5 years. (1975.0 = 0".56)
STF 23	ADS 241. Optical pair. 1836, 360°, 12".7. Comes 11.8 at 103".
STF 24	ADS 252. Fixed. w,Pb.
STT 12	ADS 434. Lambda Cas. P=640 years. Circular orbit, distance constant at 0".58. Py,Py.
STF 40	ADS 486. Fixed. Comes 12.8 at 25". y,Pb.
STT 4 Ap	ADS 513. Pi And (= H V 7 AB). w,b. Comes 11.4 at 55".
STF 1 Ap	ADS 639. Change in PA and sep. 1834, 55°, 46".4. wsh,ry?
STF 60	ADS 671. Eta Cas. P=480 years. Widening for some years yet. (1975.0 = 11".72). y,p.
STF 61	ADS 683. 65 Psc. Fixed. cpm. Py,Pb.
BU 232	ADS 684. P=218.6 years (Baize, 1962), P=150 years (Muller, 1952). Comes 10.2 at 27".
STT 20	ADS 746. 66 Psc. P=360.4 years. (1975.0 = 0".48). b,w. Comes 12.8 at 151".
STF 73	ADS 755. 36 And. P=164.7 years. (1975.0 = 0".59). y,y. Comes 11 at 160".
BU 500	ADS 768. Closing since discovery. 1878, 289°, 1".0.
BU 1099	ADS 784. P=85.79 years. (1975.0 = 0".15).
BU 302	ADS 805. 1876, 92°, 0".8.
STF 79	ADS 824. Fixed. Pb,Pb.
STT 21	ADS 862. P=450 years. (1975.0 = 0".84). Highly inclined orbit, now widening.
STF 88	ADS 899. Psi[1] Psc. Fixed. y,Pb. Comes to A, 11.2 at 94".
STT 515	ADS 940. Phi And. P=371.6 years. (1975.0 = 0".45). y,y. Slowly widening.
BU 303	ADS 955. Little motion. 1876, 284°, 0".6.
BU 235	ADS 963. Binary? 1875, 74°,0".5. C (9.0 at 59") forms STT 12 Ap. Comes to C, (11.3 at 8") forms STT 24. Comes to A, 10.4 at 43".
A 655	ADS 974. P=108.32 years. (1975.0 = 0".23). Closing.
STF 100	ADS 996. Zeta Psc. Fixed. y,Pb. B has comes 12.2 which forms BU 1129 and is probably binary. Single at times.
BU 1100	ADS 999. P=102.1 years. (1975.0 = 0".50)
STF 102	ADS 1040. 1833, 309°, 0".6. Four comites within 100". Magnitudes 8.5,10.6,11.2 and 13.1.
BU 4	ADS 1097. P=180 years. (1975.0 = 0".39). Comes 13.4 at 23".

PAIR	NOTES
STF 117 CD	ADS 1129. Psi Cas is mag 5.0 star at 283°, 25". distance decreasing. Comes 14.0 at 3" to Psi is BU 1101.
A 816	**ADS** 1226. Little motion. 1904, 313°, 0".4.
STF 138	ADS 1254. Binary of long period. w,w.
HU 1030	ADS 1264. Fixed.
STF 162	ADS 1438. Probable long period binary. 1836, 224°, 1".9. w,w. Comites 8.4 at 19", 10.2 at 137".
STT 34	ADS 1411. P=165 years (Baize,1958), P=395 years (Heintz, 1960).
STF 174	ADS 1457. 1 Ari. 1830, 170°, 2".57, Py,Pb.
HO 311	ADS 1473. Direct motion. 1890, 174°, 0".4.
STF 180	ADS 1507. Gamma Ari. 1830, 360°, 8".63. Comes 9.6 at 18", distance decreasing.
STF 183	ADS 1522. P=193 years. (1975.0 = 0".14)
STF 186	ADS 1538. P=155 years. (1975.0 = 1".29). w,w.
STF 4 Ap	ADS 1534. 56 And. Comes 11.2 at 18" to A, comes 9.4 at 204" to B.
STF 182	ADS 1531. Fixed. Comes 13.3 at 30". w,Pb.
STT 21 Ap	ADS 1563. Lambda Ari. (= H V 12 AB). Fixed. y,b.
BU 513	ADS 1598. 48 Cas. P=60.44 years. (1975.0 = 0".57). Comites 13.2 at 24", 12.6 at 51".
STF 202	ADS 1615. Alpha Psc. P=720 years. (1975.0 = 1".79). b,g.
STF 205& STT 38	ADS 1630. Gamma And. A - BC closing, 1830, 62°.4, 10".3. BC, P=61.1 years. (1975.0 = 0".54), a very eccentric orbit. y,b,b.
STF 222	ADS 1683. 59 And. Fixed. o,Pb.
STF 227	ADS 1697. Iota Tri. 1836, 80°, 3".8. y,Pb.
STF 228	ADS 1709. P=144.7 years. (1975.0 = 1".00).
STF 234	ADS 1737. P=150 years. (1975.0 = 0".89), now at widest separation. Comes 10.9 at 83".
STF 249	ADS 1795. Fixed. bsh,b.
STF 257	ADS 1833. P=209 years.
STF 271	ADS 1904. Fixed. o,Pb.
STF 93	ADS 1477. Alpha UMi. Multiple system. Visual companion is physical, 1834, 210°, 18".3. oy,bsh. Primary is a Cepheid, spectroscopic binary and has an invisible companion, P=30.5 years. Comites 12.1 at 83", 13.1 at 45".
STT 42	ADS 1938. Binary but no orbit available as yet.
STF 5 Ap	ADS 1982. 30 Ari. Fixed. y,Pb.
STF 285	ADS 2004. 1832, 178°, 1".8. oy,b.
STT 43	ADS 2034. P=475 years. (1975.0 = 1".03)
STF 299	ADS 2080. Gamma Cet. 1832, 287°, 2".6.
STF 296	ADS 2081. Theta Per. P=2720 years? w,b.
STF 305	ADS 2122. P=720 years. (1975.0 = 3".63) y,Pb. 12.6 at 88".
STF 307	ADS 2157. Eta Per. Fixed. a,b. Comes 9.9 at 66".

PAIR	NOTES
STF 314	ADS 2185. Binary. 1830, 295°, 1".5.bw,bw. Primary is a very close pair (= A 2906) and is probably binary.
BU 524	ADS 2200. 20 Per. P=31.6 years. (1975.0 = 0".21). Comes 10.0 at 15" forms STF 318.
BU 525	ADS 2253. P=433.5 years. Slowly opening. (1975.0 = 0".40)
STF 333	ADS 2257. Epsilon Ari. Binary, 1831, 190°, 0".58. w,w. Comes 12.7 at 146".
STF 331	ADS 2270. Fixed. bw,b.
STF 346	ADS 2336. 52 Ari. Binary? 1832, 264°, 0".7. w,w. Comites 10.8 at 5", 10.8 at 102" and 12.3 at 133".
STF 360	ADS 2390. P=617 years. (1975.0 = 2".50)
STT 50	ADS 2377. P=626 years. Now widening. Comes 13.2 at 27" = HJ 2172.
STF 367	ADS 2416. P=790 years. (1975.0 = 0".98).
STT 52	ADS 2436. P=330 years. (1975.0 = 0".45).
STT 53	ADS 2446. P=118.2 years. (1975.0 = 0".86).
STF 401	ADS 2582. Fixed. w,w.
STF 396	ADS 2592. Fixed. y,b. Comes 10.8 at 165".
STF 412	ADS 2616. 7 Tau. P=568 years. w,w. Comes 10.0 at 22".
STF 400	ADS 2612. P=287.7 years. (1975.0 = 1".14). Widening.
BU 533	ADS 2628. Binary? 1878, 66°, 0".4.
A 1535	ADS 2630. P=153 years (Morel,1968), P=220.5 years (Heintz,1969)
STF 422	ADS 2644. P=2100 years? 1832, 232°, 6".2. oy,bsh.
STT 37 Ap	ADS - . 1823, 95°, 41".1.
STF 430	ADS 2681. Fixed. y,Pb,Pb.
STF 427	ADS 2679. Fixed. ysh,bsh.
P III 97	ADS 2691. (= WEB AD). Fixed. ay,b. Comites 13.8 at 21" (increas 13 at 35", 10.8 at 168".
STF 452	ADS 2778. 30 Tau. Fixed. Pb,Pb.
STT 65	ADS 2799. P=62.3 years. (1975.0 = 0".72). w,w.
KUI 15	ADS - . Retrograde motion since discovery. 1937, 215°, 0".3.
STT 66	ADS 2815. Binary? 1846, 136°, 0".5.
STF 470	ADS 2850. 32 Eri. y,b. Comes 11.6 at 166".
STT 67	ADS 2867. 1847, 44°, 1".9.
STF 479	ADS 2926. Fixed.
STT 531	ADS 2995. P=706 years. (1975.0 = 1".34).
STF 495	ADS 2999. Fixed, y,bsh.
STF 460	ADS 2963. 49 Cep. P=415 years. (1975.0 = 0".80)
STT 77	ADS 3082. P=200 years. (1975.0 = 0".79). Comites 9.0 at 56" (= STT 43 Ap), 8.5 at 129".
STF 511	ADS 3098. P=254 years. (1975.0 = 0".40).
STT 48 Ap	ADS 3137. Phi Tau. (= SHJ 40). 1821, 240°, 56".8. oy,b.
STF 528	ADS 3161. Chi Tau. Fixed. Pb,g.
STT 82	ADS 3169. P=256 years. (1975.0 = 1".39). p,b.

PAIR	NOTES
HU 304	ADS 3182. 66 Tau. P=51.6 years. (1975.0 = 0".26).
STF 534	ADS 3179. 62 Tau. Fixed. ysh,b. Comes 12.0 at 111".
STF 548	ADS 3243. Fixed. y,b. Comes 10.6 at 121".
STF 550	ADS 3274. 1 Cam. Fixed. Py,Pb. Comes 11.1 at 150".
STT 86	ADS 3329. Probably binary, 1845, 79°, 0".6.
STF 567	ADS 3330. 1831, 303°, 1".4. w,w.
STF 572	ADS 3353. 1830, 210°, 3".2. y,b.
STF 11 Ap	ADS - . Sigma Tau. 1836, 193°, 431".2.
STF 577	ADS 3390. Period very uncertain. Hock, 655 years, 1966. Popovic, 1138 years, 1965. Slowly closing. y,y.
A 1013	ADS 3391. 1905, 311°, 0".5.
BU 883	ADS 3475. P=16.3 years. (1975.0 = 0".20), never wider than 0".3. Comes 14 at 17".
BU 552	ADS 3483. P=101 years. (1975.0 = 0".54). Comes 12.7 at 45".
STF 616	ADS 3572. Omega Aur. 1828, 352°, 6".5. w,Pb.
STF 623	ADS 3587. Fixed. w,Pb.
STF 627	ADS 3597. Fixed. y,g.
STF 630	ADS 3623. Fixed. y,Pb. B is A 2630 (fixed). Comes 9.5 at 130".
STT 95	ADS 3672. Binary? 1845, 344°, 0".6. w,w.
STT 98	ADS 3711. 14 Ori. P=199 years. (1975.0 = 0".72). y,b.
STF 635	ADS 3689. Opening since discovery, 1830, 281°, 0".4.
STF 645	ADS 3730. Fixed. Py,Pb. B is BU 1047 (binary)
STF 654	ADS 3797. Rho Ori. Fixed. o,b. Comes 11.8 at 182".
STT 517	ADS 3799. P=312 years. Widening. Comes 13 at 7".
STF 668	ADS 3823. Beta Ori. Fixed. b,w. Comes is a very close and very difficult binary, BU 555. Often single.
STF 653	ADS 3824. 14 Aur. Fixed. w,? Comites 11.1 at 11" and 10.4 at 184".
STF 681	ADS 3903. Fixed. y,b.
STF 634	ADS 3864. 19 Cam. 1834, 349°, 34". Due to pm of A.
STF 696	ADS 3962. 23 Ori. Fixed. Py,b.
STF 701	ADS 3978. Fixed. w,Pb.
WNC 2& A 847	ADS 3991. A - BC is WNC 2. Binary, 1866, 171°, 1".4. w,y. BC is also binary, P=24.68 or 49.36 years, orbit highly inclined to the line of sight.
DA 5	ADS 4002. Eta Ori. Little motion, 1849, 87°, 1".0. w,b. Comes 9.4 at 115".
HO 226	ADS 4032. Widening, 1887, 230°, 0".5. Comes 10.5 at 25".
STF 716	ADS 4068. 118 Tau. 1829, 197°, 4".9. ysh,bsh. Comes 11.6 at 141".
STF 728	ADS 4115. 32 Ori. P=586 years (Siegrist , 1951) but van den Bos (1956) and Cester (1964) think motion is linear. w,w.
STF 729	ADS 4123. 33 Ori. Fixed. Pb,b. Comes 13.4 at 95".

PAIR	NOTES
STF 14 Ap	ADS 4134. Delta Ori. Fixed. w,w. Comes 14.0 to A at 33" forms BU 558.
STF 718	ADS 4119. Fixed. Pb,w. Comes 9.2 at 119".
BU 1267	ADS 4166. 1892, 218°, 0".8.
STF 738	ADS 4179. Lambda Ori. Fixed. w,b. Comites 11.2 at 29", 78".
STF 748	ADS 4186. Theta^1 Ori. "The Trapezium". Fixed.
STF 752	ADS 4193. Iota Ori. Fixed. bw,b. Enmeshed in nebulosity.
STF 742	ADS 4200. 1837, 251°, 3".3. o,o.
STF 749	ADS 4208. Probable binary, 1829, 23°, 0".7. w,w. Pair nearby, 10.4, 10.9, 290°, 4".4.
STF 753&	ADS 4229. 26 Aur. AB - C is STF 753. Fixed. w,b. AB is BU 1240.
BU 1240	P=53.2 years, (1975.0 = 0".06). Comes to A, 11.5 at 33".
STT 112	ADS 4243. Probably binary, 1848, 85°, 0".6.
STF 766	ADS 4256. Fixed. Py,b.
STF 774	ADS 4263. Zeta Ori. P=1500 years? 1836, 151°, 2".6. Comes to A, 10.0 at 58".
BU 1007	ADS 4265. 126 Tau. P=78.5 years. (1975.0 = 0".23).
STF 764	ADS 4262. Fixed. w,b.
STT 115	ADS 4323. 1847, 123°, 0".8. Comes 11.0 at 93".
BU 560	ADS 4371. 1877, 208°, 0".9.
STF 795	ADS 4390. 52 Ori. Likely long period binary. o,o.
STT 118	ADS 4392. Fixed. Comes 8,6 at 76" forms STT 67 Ap.
STF 3115	ADS 4376. Closing since discovery. 1831, 36°, 1".7.
STF 802	ADS 4456. Fixed. w,w. Comes 10.6 at 192".
A 1570	ADS 4531. Fixed.
STT 545	ADS 4566. Theta Aur. Long period binary. 1871, 6°, 2".2. Comites 10.7 at 50", 9.2 at 131", distances increasing.
STT 121	ADS 4603. Closing since discovery. 1843, 191°, 0".4.
STF 855	ADS 4749. Fixed. w,b,w.
STT 134	ADS 4744. Fixed. o,b. On nf edge of M35.
STF 845	ADS 4773. 41 Aur. 1830, 353°, 8".0. w,Pb.
KUI 24	ADS - . 1934, 137°, 0".3.
STF 877	ADS 4840. Fixed.
BU 1008	ADS 4841. Eta Gem. Primary is a M3 variable. 1882, 301°, 1".0.
STF 872	ADS 4849. 1828, 217°, 11".3. w,Pb. Comites 11.4 at 202", 11.5 at 174".
STF 881	ADS 4950. 4 Lyn. Binary. 1830, 90°, 0".8. Py,b. Comites 12.9 at 26", 11 at 100".
STF 919	ADS 5107. Beta Mon. Fixed. w,w,w. Comes to A, 12.2 at 25" forms BU 570.
STF 924	ADS 5166. 20 Gem. Fixed. Py,Pb.
STF 932	ADS 5197. 1830, 342°. 2".4.
STF 928	ADS 5191. Fixed. ysh,bsh. Comes 11.0 at 129".
STT 149	ADS 5234. P=114.8 years. (1975.0 = 0".56)

PAIR	NOTES
STT 152	ADS 5289. 54 Aur. Fixed. y,b.
STF 945	ADS 5296. Closing since discovery. 1830, 249°, 1".1.
STF 946	ADS 5368. Fixed.
AGC 1	ADS 5423. Alpha CMa. P=50.09 years, (1975.0 = 11".18). Comes is the brightest known white dwarf. Occasionally thought double. bw,w. Comes 14.0 at 32".
STF 948	ADS 5400. 12 Lyn. A triple system. AB has a period of 699 years. (1975.0 = 1".69).AC, 1831, 308°, 8".7 - little change. y,Pb,b. Comes 10.5 at 170".
STT 78 Ap	ADS 5425. Psi^5 Aur (= SHJ 75). 1823, 17°, 55".4. o,b.
STT 156	ADS 5447. P=1058 years. (1975.0 = 0".54)
STF 958	ADS 5436. Fixed. y,rsh. Comes 11.2 at 164".
STF 963	ADS 5514. 14 Lyn. P=480 years. (1975.0 = 0".42). Comes 11.1 at 181".
STF 982	ADS 5559. 38 Gem. P=3190 years? ysh,bsh. Comes 10.3 at 112".
STF 981	ADS 5570. Closing since discovery. 1831, 149°, 3".7.
STT 159	ADS 5586. 15 Lyn. 1844, 323°, 0".5. y,b. Comites 12.4 at 29", 8.9 at 197".
A 1575	ADS 5704. Fixed.
STT 81 Ap	ADS 5742. Zeta Gem. (= SHJ 77AC)Primary is a Cepheid variable. y,b. Comites 10.5 at 87", 12.0 at 80" and 12.9 at 27".
STF 1009	ADS 5746. Widening, 1830, 159°, 2".9. Pb,Pb. Comes 11.0 to B at 179".
STF 1033	ADS 5896. Fixed. y,b.
STT 170	ADS 5958. Widening since discovery. 1844, 133°, 1".0. Py,Py.
STF 1066	ADS 5983. Delta Gem. P=1200 years? y,Pb.
STF 1083	ADS 6060. No change. y,g.
STF 1093	ADS 6117. P=381 years. (1975.0 = 0".75). Widening.
STF 1108	ADS 6160. Fixed. y,b.
STF 1110	ADS 6175. Alpha Gem. P=420 years. (1975.0 = 1".96). Now just past periastron. Castor C, 8.8 at 73" is part of the system. All three are spectroscopic bins.
STT 175	ADS 6185. 1847, 334°, 0".5. Distance decreasing. or,or. Comes 9.3 at 81".
STT 174	ADS 6191. 1851, 85°, 2".0. Distance increasing? y,b.
STF 1126	ADS 6263. Probably binary. 1829, 132°, 1".5. bsh,bsh. Comes 10.6 at 44".
STT 177	ADS 6276. P=170 years or 201 years.
STT 179	ADS 6321. Kappa Gem. Probable long period binary. 1853, 233°, 6".2. Py,Pb.
STF 1138	ADS 6348. 2 Pup. Fixed. y,l. Comes 10.4 at 101".
STF 1140	ADS 6376. Fixed. Py,Pb.
STT 186	ADS 6538. Fixed. w,w

PAIR	NOTES
BU 581	ADS 6554. P=45 years. (1975.0 = 0".42). Comes 10.5 at 4".8 is a member of the system.
STF 1177	ADS 6569. Fixed. w,b.
STF 1187	ADS 6623. Binary? 1829, 71°, 1".6.
STF 1196	ADS 6650. Zeta Cnc. AB has P=59.7 years (1975.0 = 0".96), BC, P=1150 years? y,y,Pb. Irregularities in the motion of C reveal a dark component c. Period of Cc = 17.5 years.
HU 1123	ADS 6681. Fixed.
STF 1216	ADS 6762. P=435 years. (1975.0 = 0".60). Widening slowly.
STT 93 Ap	ADS - . 41 Lyn. (= S 565). 1824, 165°, 73". y,Pb.
STF 1224	ADS 6811. 24 Cnc. (Not Upsilon[1] Cnc as given in ADS and Webb). 1830, 37°, 5".8. w,bsh. B is A 1746, P=21.82 years. (1975.0 = 0".16).
STF 1223	ADS 6815. Phi[2] Cnc. 1829, 212°, 4".6. w,w.
STF 1245	ADS 6886. Fixed. y,l. Comites 10.7 at 93", 12.0 at 117" and 8.8 at 122".
STF 1268	ADS 6988. Iota Cnc. Fixed. o,Pb.
STF 1273	ADS 6993.(AC) Epsilon Hydrae. P=890 years. y,g. AB is a rapid visual binary, P=15.05 years. (1975.0 = 0".15). A and B are also spectroscopic binaries. A comes 12.5 at 19" is attached to the system.
STF 1275	ADS 7033. Fixed. Comes 12.6 at 40" forms HJ 2466 AC.
A 1584	ADS 7054. P=74.1 years. (1975.0 = 0".40). Closing.
STF 1291	ADS 7071. 57(Iota[2]) Cnc. 1829, 333°, 1".5. y,y. Comes 9.2 at 56".
KUI 37	ADS - . 10 UMa. P=21.85 years. (1975.0 = 0".40). Comites 10.8 at 142", 9.8 at 127" and 10.5 at 225". Distances changing due to large pm of AB.
STF 1300	ADS 7139. 1830, 210°, 4".1. Distance increasing.
STF 1298	ADS 7137. 66 Cnc. Fixed. w,Pb. Comes 10.8 at 187".
STF 1311	ADS 7187. Fixed. ysh,w. Comes 12.6 at 28" forms HO 644 AC.
STF 1306	ADS 7203. Sigma[2] UMa. P=1067 years. (1975.0 = 3".03).
STF 1321	ADS 7251. P=687 years. o,o.
STF 3121	ADS 7284. P=34.2 years. (1975.0 = 0".56). Closing rapidly.
STF 1333	ADS 7286. 1828, 39°, 1".4. ysh, bsh.
STF 1334	ADS 7292. 38 Lyn. 1829, 240°, 2".7. Retrograde motion. w,b. Comites 10.8 at 88", 10.7 at 178".
STF 1338	ADS 7307. P=389 years (Arend, 1953), P=220 years (Guntzel-Lingner, 1953). w,w. Comes 11.4 at 143".
STF 1348	ADS 7352. 1831, 334°, 1".1. Pb,Pb.
STF 1356	ADS 7390. Omega Leo. P=116.85 years. (1975.0 = 0".49).
STF 1360	ADS 7406. Fixed. Pb,g. Comites 12.3 at 85" and 11.2 at 158".
STF 1351	ADS 7402. 23 UMa. Fixed. y,bsh. Comes 10.5 at 100".
STT 101 Ap	ADS 7416. 6 Leo. (= SHJ 107). Fixed. o,Pb.
STF 1374	ADS 7477. 1838, 275°, 3".3.
A.C.5	ADS 7555. Gamma Sex. P=75.6 years. (1975.0 = 0".44). Opening. Comes 12.1 at 36" forms HJ 4256 AC.

PAIR	NOTES
STF 1399	ADS 7589. Fixed. Py,g.
STT 215	ADS 7704. P=552 years. w,w.
STF 1415	ADS 7705. Fixed. Comes 10.6 at 150".
STF 1424	ADS 7724. Gamma Leo. P=619 years. (1975.0 = 4".28). o,o. Distant comites 9.6, 10.0.
STF 1426	ADS 7730. 1832, 257°, 0".6. Direct motion. Comes 9.3 at 8".
STF 1428	ADS 7762. 1831, 84°, 3".8.
HU 879	ADS 7780. Beta LMi. P=37.90 years. (1975.0 = 0".55).
STF 1439	ADS 7802. 1829, 131°, 2".0.
STF 1450	ADS 7837. 49 Leo. Fixed. w,Pb.
STF 1457	ADS 7864. 1829, 288°, 0".7.
STT 228	ADS 7926. 1851, 196°, 0".5.
STT 229	ADS 7929. Long period binary. 1846, 347°, 0".7.
STF 1487	ADS 7979. 54 Leo. 1830, 103°, 6".2. w,Pb.
BU 1077	ADS 8035. Alpha UMa. P=44.4 years. Near apastron.
HO 378	ADS 8047. 1891, 219, 0".4. Widening since discovery.
STF 1517	ADS 8094. 1829, 288°, 1".0. Nature of motion rather uncertain. Quadrant determination difficult.
STT 232	ADS 8102. Fixed.
STF 1516	ADS 8100. Optical pair due to large motion of A. Comes to A, 11.2 at 7" forms STT 539 and is probably a physical component.
STF 1523	ADS 8119. Xi UMa. P=59.84 years. (1975.0 = 3".03). Py,Py. Both stars are spectroscopic binaries.
STF 1524	ADS 8123. Nu UMa. Fixed. y,b.
STF 1527	ADS 8128. P=1150 years? 1829, 10°, 3".9. ysh,g.
STF 1536	ADS 8148. Iota Leo. P=192 years. (1975.0 = 1".15).
STF 1543	ADS 8175. 57 UMa. 1831, 11°, 5".4. w,r? Comites 11.6 at 217", 7.8 at 346" and two to 7.8, 10.2 at 98" and 10.3 at 129".
STT 234	ADS 8189. P=86.4 years. (1975.0 = 0".17).
STF 1547	ADS 8196. 88 Leo. 1782, 318°, 14".6. w,Pb.
STT 235	ADS 8197. P=71.9 years. Closing rapidly. w,w.
STF 1552	ADS 8220. 90 Leo. Fixed. w,w. Comes 8.8 at 63".
STF 1555	ADS 8231. Highly inclined orbit? 1829, 339°, 1".2. Comes 10.2 at 21" forms HJ 503 AC.
STF 1559	ADS 8249. Fixed.
STF 1561	ADS 8250. 1831, 266°, 10".5. A has pm of 0".6 annually. Comites 8.6 at 90", 9.5 at 121" and 12.4 at 91".
STT 237	ADS 8252. 1845, 287°, 0".7.
BU 794	ADS 8337. P=77.2 years. (1975.0 = 0".34). Comites 14.0 at 5".6, 13.3 at 27".
A 1088	ADS 8387. Direct motion. 1905, 223°, 0".3.
STF 1596	ADS 8406. 2 Com. Fixed. Py,Pb.
STF 3123	ADS 8419. P=120 years. (1975.0 = 0".15). Comes 15.6 at 3" is probably physically attached. Comes 8.0 at 26" is double (0".2).

PAIR	NOTES
STF 1604	ADS 8440. 1831, 93°, 12".0 (AB), 1831, 97°, 58".0 (AC)
STF 1606	ADS 8446. Long period binary. 1831, 349°, 1".4.
STF 1622	ADS 8489. 2 CVn. Fixed. o,b.
STF 1627	ADS 8505. Fixed. w,w.
STF 1633	ADS 8519. Fixed. bw,bw.
STT 249	ADS 8535. Closing since discovery, 1853, 315°, 0".5. Comes 11.2 at 13".
STF 1639	ADS 8539. P=678 years. (1975.0 = 1".44).
STT 250	ADS 8540. 1845, 331°, 0".4. Little motion.
STF 1643	ADS 8553. P=2200 years?
STF 1647	ADS 8575. P=1140 years?
STF 1657	ADS 8600. 24 Com. Fixed. Py,bsh.
STF 1661	ADS 8606. 1828, 226°, 2".6.
STF 1670	ADS 8630. Gamma Vir. P=171.4 years. (1975.0 = 4".25). Highly eccentric orbit. Closing to 0".4 in AD 2008. w,w. Comes 12.2 at 124".
STF 1678	ADS 8659. 1832, 212°, 32".6. Pb,oy.
STF 1687	ADS 8695. 35 Com. P=674 years. (1975.0 = 0".83). y,y. Comes 9.1 at 29", b.
STF 1689	ADS 8704. 1827, 198°, 28".7. Dy,g.
STF 1692	ADS 8706. Alpha CVn. Fixed. w,w. Primary is a magnetic variable.
STF 1695	ADS 8710. 1832, 289°, 3".3. Comes 10.4 at 124".
STT 256	ADS 8708. 1848, 57°, 0".7.
STF 1699	ADS 8721. 1830, 1°, 1".5. Small change in angle.
BU 1082	ADS 8739. 78 UMa. P=115.7 years. (1975.0 = 1".27). Widening.
BU 929	ADS 8759. 48 Vir. 1879, 229°, 0".5. Widening.
STF 1728	ADS 8804. 42 Com. P=25.87 years. (1975.0 = 0".04). Orbit almost edge-on to the line of sight. Now just past conjunction. Comes 10.2 at 90" (decreasing due to pm of AB).
STT 261	ADS 8814. 1843, 359°, 0".6. y,Pb.
STF 1734	ADS 8864. 1830, 198°, 0".7. Distance increasing. w,w.
STF 1744	ADS 8891. Zeta UMa. AB is a long period binary, C (Alcor) at 709" has the same pm as AB. A and B are also spectroscopic binaries. w,w.
STF 1742	ADS 8890. Fixed. bw,bw.
STT 269	ADS 8939. P=54.8 years. (1975.0 = 0".26)
STF 1757	ADS 8949. P=334 years. (1975.0 = 2".29). Now closing. Comes 11.6 at 45".
STF 1768	ADS 8974. 25 CVn. P=240 years. Eccentric orbit. Comes 8.7 at 217". Py,w.
STF 1770	ADS 8979. Fixed. o,b.
BU 612	ADS 8987. P=22.35 years. (1975.0 = 0".09). Comes 11.0 at 125" distance decreasing.

PAIR	NOTES
STF 1781	ADS 9019. P=300 years.
STF 1785	ADS 9031. P=155 years. (1975.0 = 3".34). Py,bw.
STT 276	ADS 9121. 1845, 196°, 0".6. Comes at 9".7. IDS gives magnitude as 10.0 and 10.8.
STT 278	ADS 9159. P-203 years. (1975.0 - 0".26).
STT 277	ADS 9158. 1845, 334°, 0".4. Comes 9.3 at 14".2 forms STF 1812. Further comes 11.8 at 73".
STF 1820	ADS 9167. P=810 years? Circular orbit? 1831, 47°, 2".4.
STF 1821	ADS 9173. Kappa^2 Boo. 1832, 238°, 12".6. w,bsh.
STF 1816	ADS 9174. Closing. 1831, 80°, 1".9.
STF 1817	ADS 9177. Closing. 1832, 7°, 1".6.
STF 1834	ADS 9229. P=321 years. (1975.0 = 1".19). Orbit of high eccentricity and inclination. w,w.
A 570	ADS 9301. P=30.0 years. (1975.0 = 0".20).
STF 1863	ADS 9329. 1830, 110°, 0".6.
STF 1864	ADS 9338. Pi Boo. 1830, 99°, 5".8. w,w. Comes 10.0 at 128".
STF 1865	ADS 9343. Zeta Boo. P=123.4 years. Very eccentric orbit. w,w. Comes 10.9 at 99" is H VI 104.
STF 1871	ADS 9350. 1829, 283°, 1".8. Widening. w,w.
STF 1877	ADS 9372. Epsilon Boo. 1829, 321°, 2".6. y,b. Comes 12.0 at 177".
STT 285	ADS 9378. P=88.4 years. (1975.0 = 0".17). Just past periastron.
STF 1884	ADS 9389. 1829, 55°, 1".7.
BU 106	ADS 9396. Mu Lib. Opening, 1875, 335°, 1".4. Comites 13.9 at 25", 12.5 at 27".
STF 1890	ADS 9406. 39 Boo. 1830, 44°, 3".7. w,w.
STT 287	ADS 9418. P=400 years. (1975.0 = 1".10)
STF 1888	ADS 9413. Xi Boo. P=151.5 years. (1975.0 = 7".16). or,b. Comites 12.6 at 67", 9.6 at 149".
STT 288	ADS 9425. P=215.4 years. (1975.0 = 1".34). Pb,Pb.
BU 348	ADS 9480. 2 Ser. 1875, 115°, 0".5.
STF 1909	ADS 9494. 44 Boo. P=246.2 years. (1975.0 = 0".54). Very highly inclined orbit. B is a W UMa variable.
A 1116	ADS 9530. 1905, 21°, 0".4.
STF 1932	ADS 9578. P=203 years. (1975.0 = 1".30). B has a dark companion period 50 years.
BU 32	ADS 9596. 6 Ser. 1875, 13°, 2".3.
STF 1937	ADS 9617. Eta CrB. P=41.56 years. (1975.0 = 0".54). y,y. Comites 11.0 at 215", 12.6 at 58".
STF 1938	ADS 9626. Mu^2 Boo. P=260 years. (1975.0 = 2".14). w,w. Mu^1 is 108" distant, magnitude 4.5.
HU 149	ADS 9628. 1900, 296°, 0".2.
STT 296	ADS 9639. 1845, 328°, 1".5. Comes 12.5 at 67".

PAIR	NOTES
STF 1956	ADS 9694. 1831, 41°, 2".7. Closing with little change in PA.
STF 1954	ADS 9701. Delta Ser. 1819, 202°, 3".4. Long period binary. ysh,bsh.
STT 298	ADS 9716. P=55.9 years (1975.0 = 0".97). w,w. Comes 7.9 at 121" is apparently physical.
STF 1963	ADS 9727. 1829, 291°, 4".2. w,w. Comes 13.4 at 31".
STF 1965	ADS 9737. Zeta CrB. 1822, 299°, 6".2. Long period binary. bsh,Pb.
STF 1989	ADS 9769. Pi^2 UMi. P=150.8 years. (1975.0 = 0".61). Highly eccentric orbit.
STF 1969	ADS 9756. Eccentric orbit of long period now past periastron. 1831, 43°, 1".5.
HU 580	ADS 9744. Iota Ser. P=11.07 or 22.14 years. Quadrant determination difficult because components are equally bright and very close at all times. Comites 13.4 at 143" and 12.6 at 151".
STF 1967	ADS 9757. Gamma CrB. P=91.0 years. (1975.0 = 0".25). Highly inclined orbit.
BU 619	ADS 9758. Fixed? w,w.
STT 303	ADS 9880. 1846, 111°, 0".6. Widening since discovery. w,w.
STF 1998	ADS 9909. Xi Sco. Period of AB = 45.69 years. (1975.0 = 1".24). y,y. C also belongs to the system - 1825, 79°, 6".8 (bsh). STF 1999 at 281" has similiar pm and is associated.
STF 1999	ADS 9910. 1831, 102°, 10".5. Distance increasing slightly.
BU 812	ADS 9925. 1881, 127°, 0".9. Closing.
BU 120	ADS 9951. Nu Sco. A double-double. AB is fixed, 1876, 3°, 0".9. CD is widening, 1846, 39°, 1".1. AC is H V 6.
STF 2021	ADS 9969. 49 Ser. P=1350 years? 1829, 316°, 3".2. y,b. Comes 10.6 at 236".
STF 2032	ADS 9979. Sigma CrB. P=1000 years. (1975.0 = 6".58). y,p. Two comes 13.3, 10.8 whose distances are changing rapidly due to the pm of AB.
STT 309	ADS 10006. 1846, 236°, 0".5. Probably binary.
STF 2054	ADS 10052. 1832, 7°, 0".9. Slow binary? w,w.
STF 2049	ADS 10070. Binary. w,w.
STF 2052	ADS 10075. P=236 years. (1975.0 = 1".28).w,w. Comes 11.1 at 14
STF 2055	ADS 10087. Lambda Oph. P=129.9 years. (1975.0 = 1".25). yg,yg. Comes 11.0 at 119".
STT 313	ADS 10111. 1847, 162°, 0".8. Change mainly in angle.
STF 2078	ADS 10129. 17 Dra. 1831, 116°, 3".7. w,w. C, mag 5.6 at 90" forms STT 30 Ap and has comes 11.6 at 120".
STF 2084	ADS 10157. Zeta Her. P=34.385 years. (1975.0 = 1".15).
STF 2091	ADS 10169. 1830, 302°, 1".3. Probable binary.
HU 664	ADS 10189. Fixed.
D 15	ADS 10188. P=120 years. Now beginning to close.
STF 2094	ADS 10184. 1831, 83°, 1".6. w,w. Comes 11.0 at 25".

PAIR	NOTES
STF 2106	ADS 10229. P=1080 years. (1975.0 = 0".50). Now opening.
STT 315	ADS 10230. 21 Oph. 1844, 173°, 0".9. Closing. w,w.
STF 2107	ADS 10235. P=261.8 years. (1975.0 = 1".35). Pb,Pb.
STF 2118	ADS 10279. 20 Dra. P=729 years. (1975.0 = 1".22) w,w.
STF 2114	ADS 10312. 1830, 136°, 1".3. Probable binary. y,l.
STF 2130	ADS 10345. Mu Dra. P=482 years. (1975.0 = 1".94). Now slowly closing. Comes 13.8 at 13" forms BU 1088 and appears physical.
KUI 79	ADS - . P=12.98 years. (1975.0 = 0".94). Will now close rapidly.
STF 2140	ADS 10418. Alpha Her. 1829, 119°, 4".7. Probably physical. o,g. Comes 11.1 at 80".
STF 2161	ADS 10526. Rho Her. 1830, 307°, 3".6. Probably physical. w,w. Comes 13.3 at 119".
BU 1201	ADS 10573. Fixed.
HO 417	ADS 10589. 1892, 151°, 0".4.
STF 2173	ADS 10598. P=46.1 years. (1975.0 = 0".18). Orbit of high inclination.
STF 2207	ADS 10690. 1832, 128°, 1".1. Probably binary.
STF 2199	ADS 10699. 1830, 116°, 1".7. Retrograde motion,
STF 2218	ADS 10728. 1836, 355°, 2".5. Angle decreasing.
STF 2203	ADS 10722. 1830, 334°, 0".7. Retrograde motion.
A.C.7	ADS 10786. BC is A.C.7, P=43.2 years. (1975.0 = 0".56) (Mu[2] Her). Mu[1] Her is 3.5 at 35" and forms STF 2220 with BC. It has a comes 11.2 at 256".
STF 2215	ADS 10795. 1831, 311°, 0".8. Retrograde motion.
STT 337	ADS 10828. P=500 years. (1975.0 = 0".30)
STF 2242	ADS 10849. Fixed.
STT 338	ADS 10850. 1845, 44°, 0".7. Slow binary. w,w. Comites 12.9 at 28", 9.9 at 95".
STF 2264	ADS 10993. 95 Her. 1829, 262°, 6".1. w,w.
STF 2267	ADS 11001. 1830, 234°, 1".4. Closing with increasing angle.
STF 2262	ADS 11005. Tau Oph, P=280 years. (1975.0 = 1".89).y,y. Comes 9.3 at 100".
STF 2272	ADS 11046. 70 Oph. P=87.85 years. (1975.0 = 1".88). y,p. Many comites given in the IDS.
STT 341	ADS 11060. P=20.0 years. (1975.0 = 0".35). Orbit of extremely high eccentricity and high inclination. Comites mag 9.6 at 28", 38", 63" and 108". Also 8.6 at 136".
HU 1186	ADS 11071. P=101.5 years. (1975.0 = 0".38). Now near its widest separation.
STF 2281	ADS 11111. 73 Oph. P=270 years. (1975.0 = 0".32). Comes 12.6 at 68".
HU 674	ADS 11128. 1904, 279°, 0".5. Retrograde motion.
STF 2289	ADS 11123. P=3000 years? w,c.
STF 2294	ADS 11186. P=278 years. (1975.0 = 1".00). Orbit of high eccentricity.

PAIR	NOTES
STT 353	ADS 11311. Phi Dra. 1856, 64°, 0".6.
A.C.11	ADS 11324. P=240 years. (1975.0 = 0".78).
STF 2315	ADS 11334. P=775 years. (1975.0 = 0".66). Comes 13.5 at 45".
STT 351& HU 66	ADS 11344. AC is STT 351, 1846, 25°, 0".5, widening slowly. AB is HU 66, 1898, 310°, 0".3 - retrograde motion. One of the closest triples known,
STF 2339	ADS 11454. 1830, 272°, 2".3. Primary is HU 322, a very close pair in orbital motion.
A 1377	ADS 11468. P=185 years or 338 years. Comes 8.8 at 26" forms STF 2348.
STT 359	ADS 11479. P=211 years. (1975.0 = 0".58). w,w.
STT 358	ADS 11483. P=292 years. (1975.0 = 1".63).
STT 357	ADS 11484. P=256 years. (1975.0 = 0".32).
STF 2351	ADS 11500. 1830, 340°, 5".2. Fixed.
STF 2384	ADS 11568. P=150 years. (1975.0 = 0".95). Comes 12.5 at 113".
STF 2368	ADS 11558. 1831, 331°, 2".0. Comes 10.7 at 37".
HO 437	ADS 11566. 1892, 116°, 0".4. Comites 10.7 at 40", 13.9 at 23" and 11.2 at 4" to 10.7. (distance increasing).
STF 2367	ADS 11579. P=90 years. (1975.0 = 0".13). Very eccentric orbit. Comites 8.6 at 14", 11.9 at 23", 10.9 at 149".
STF 2398	ADS 11632. P=453 years. (1975.0 = 14".48). Comes 12.8 at 200" (distance increasing due to large pm of AB).
STF 2382	ADS 11635. Epsilon 1 Lyr. P=1165 years. (1975.0 = 2".71). Linked physically with STF 2383. w,w.
STF 2383	ADS 11635. Epsilon 2 Lyr. P=585 years. (1975.0 = 2".32). Several comites to each pair.
STF 2415	ADS 11816. 1831, 299°, 2".0.
BU 648	ADS 11871. P=61.2 years. (1975.0 = 0".69). Comites 12.1 at 55", 12.2 at 91".
STF 2422	ADS 11869. 1832, 106°, 0".9. Long period binary.
STF 2438	ADS 11897. P=259 years. (1975.0 = 0".87). Eccentric orbit.
STF 2437	ADS 11956. 1830, 81°, 1".1. Closing since discovery.
SHJ 286	ADS 12007. 15 Aql. 1823, 207°, 35".6. o,l.
A 863	ADS 12108. 1904, 123°, 0".4.
STF 2481& SE 2	ADS 12145. A - BC is STF 2481, 1830, 234°, 3".8. BC is SE 2, P=63.1 years. (1975.0 = 0".42).
STF 2486	ADS 12169. 1832, 225°, 10".5. ysh,ysh. Comites 13.3 at 25" (dist. decreasing), 11.4 at 180"(increasing).
BU 139	ADS 12160. Fixed. Comes 7.9 at 113" forms STT 177 Ap. Also comites 9.5 at 29" and 12.7 at 28".
STT 371	ADS 12239. 1846, 154°, 0".8. w,w. Comes 8.6 at 48" (b).
STF 2491	ADS 12246. 1828, 207°, 1".1.
STF 2509	ADS 12296. 1832, 353°, 0".5. Becoming wider.
BU 1129	ADS 12366. 1889, 344°, 0".3.
STF 2525	ADS 12447. P=990 years. (1975.0 = 1".77). Eccentric orbit,

PAIR	NOTES
STT 377	ADS 12667. 1842, 51°, 0".9. Comes 10.1 at 25".
STF 2556	ADS 12752. P=256 years. (1975.0 = 0".22).
STF 2574	ADS 12803. 1832, 129°, 1".0.
STT 380	ADS 12808. Chi Aql. Fixed. Comes 12.3 at 82". Also 10.3 at 140" forms J 1858. Nearby pair 11.0, 11.5, 317°, 9".4 is HLM 26.
STT 383	ADS 12831. 1845, 27°, 0".9.
STT 384	ADS 12851. Fixed. w,w.
STF 2579	ADS 12880. Delta Cyg. P=537 years. (1975.0 = 2".18). bw,Pb. Comes 12.0 at 66".
STF 2576	ADS 12889. P=224.7 years. (1975.0 = 1".85). w,w.
STT 386	ADS 12965. 1846, 78°, 1".0.
STF 2603	ADS 13007. Epsilon Dra. 1832, 354°, 2".8. Py,bsh.
STF 2583	ADS 12962. Pi Aql. 1829, 121°, 1".5. Pb,c. Comes 12.2 at 34".
STT 387	ADS 12972. P=156.5 years. (1975.0 = 0".59).
A.G.C.11	ADS 12973. P=22.80 years. (1975.0 = 0".18). C (8.8 at 9") forms STF 2585.
HO 581	ADS 13125. P=25.69 years. (1975.0 = 0".31).
STF 2605	ADS 13148. Psi Cyg. 1831, 185°, 3".3. Py,Pb. Comites 13.6 at 21" and 10.2 at 165".
STF 2606	ADS 13196. 1832, 131°, 1".2.
STT 395	ADS 13277. 16 Vul. 1844, 79°, 0".6. Py,Py.
STF 2624	ADS 13312. 1830, 179°, 2".0. w,w. Comites 9.1 at 43" and 11.0 at 29".
STF 2626	ADS 13329. Fixed.
STF 2642	ADS 13392. 1832, 165°, 2".4. Comes 12.5 at 73" (dist dec.)
STF 2652	ADS 13449. 1832, 280°, 0".3. Long period binary.
STT 400	ADS 13461. P=86.2 years. (1975.0 = 0".24).
STT 403	ADS 13572. 1848, 173°, 0".8. Star C, 9.0 at 11".6 forms STF 2657.
STF 2671	ADS 13692. 1831, 341°, 3".0. w,w. Comes 12.5 at 84".
STT 406	ADS 13723. P=96 years. (1975.0 = 0".52).
D 22	ADS 13847. 1875, 140°, 2".8. Comes 13.8 at 19" (dist. dec.)
HO 130	ADS 13856. Fixed.
BU 151	ADS 14073. Beta Del. P=26.6 years. (1975.0 = 0".61). Comes 11.0 at 39" forms STF 2704. Also 13.1 at 23".
STF 2705	ADS 14078. Fixed. Py,Pb. Comites 11.2 at 183", 12.9 at 5.4"
STT 533	ADS 14101. Kappa Del. 1852, 11°, 10".4. Rectilinear motion.
STT 410	ADS 14126. 1850, 23°, 0".6. bw,bw. Comes 8.6 at 69".
STF 2716	ADS 14158. 49 Cyg. Fixed. Py,Pb. Comes 11.8 at 68".
STF 2723	ADS 14233. 1831, 86°, 1".5. Round in 1958.
STF 2726	ADS 14259. 52 Cyg. 1830, 57°, 6".6. o,Pb.
BU 676	ADS 14274. 1878, 321°, 37".7. Optical pair. Gamma Cyg.
STF 2725	ADS 14270. 1829, 358°, 4".2. w,w.
STF 2727	ADS 14279. Gamma Del. 1830, 274°, 11".9. y,g.

PAIR	NOTES
STT 413	ADS 14296. Lambda Cyg. P=391 years. (1975.0 = 0".83). w,Pb. Comes 9.9 at 85" forms S 765.
BU 268	ADS 14306. 1875, 221°, 0".4.
BU 155	ADS 14370. 1876, 25°, 0".6. Comites 12.0 at 14", 10.9 at 195". Main pair w,b.
HO 144	ADS 14379. 1886, 168°, 0".4. Rapid change.
HO 146	ADS 14404. 1886, 56°, 0".4.
STT 418	ADS 14421. 1842, 302°, 0".6. w,w.
STF 2741	ADS 14504. 1831, 36°, 1".9. Py,Pb. Comes 10.5 at 140".
STF 2737	ADS 14499. Epsilon Equ. P=101.4 years. (1975.0 = 1".06). Orbit very highly inclined. Star C is physical, mag 7.1, (1833, 78°, 10".8) - orbit hyperbolic? Comes 12.4 at 65".
STF 2742	ADS 14556. 2 Equ. 1831, 225°, 2".6. Py,Pb.
STF 2751	ADS 14575. 1831, 344°, 1".9.
STF 2744	ADS 14573. 1830, 190°, 1".5. w,w. Comes 12.8 at 89".
STF 2758	ADS 14636. 61 Cyg. P=653 years. (1975.0 = 28".69). y,y. Dark companion orbits A, P=4.90 years, probably planetary.
STF 2760	ADS 14645. 1829, 223°, 13".7. Change due to pm. Comes 9.8 at 60".
STF 2780	ADS 14749. 1831, 229°, 1".1. w,w. Comes 8.7 at 121".
H I 48	ADS 14783. P=84.4 years. (1975.0 = 0".77).
STT 432	ADS 14778. 1847, 130°, 1".2. y,b,
STT 535	ADS 14773. Delta Equ. P=5.70 years. (1975.0 = 0".05). A highly inclined orbit. Comes 9.5 at 60" forms STF 2777 - distance increasing due to pm of AB.
HO 286	ADS 14859. Rapid binary, often single. Period uncertain.
STT 437	ADS 14889. 1845, 68°, 1".4. y,Pb. Comes 11.2 at 86".
STT 435	ADS 14894. 1848, 203°, 0".6.
HO 157	ADS 14928. Fixed.
STF 2799	ADS 15007. 1831, 333°, 1".4. Retrograde motion. bw,bw. Comes 9.3 at 130".
STF 2804	ADS 15076. 1828, 314°, 2".9. The PA in Webb is a misprint Burnham's measure should be 335°.7. y,y. Comes 11.6 at 97".
STT 445	ADS 15177. Fixed.
BU 1212	ADS 15176. 24 Aqr. P=48.7 years. (1975.0 = 0".19). Orbit of high eccentricity. Comes 10.9 at 36" (distance dec. due to pm of AB).
HO 166	ADS 15267. P=80.5 years. (1975.0 = 0".39). Near apastron.
STF 2822	ADS 15270. Mu Cyg. P=508 years. (1975.0 = 1".86). Orbit of high inclination. Mu^2 (6.9) is at 196", distance dec. Comes 13.3 to Mu^2 forms ES 521. Main pair ysh,bw.
COU 14	ADS - . 13 Peg. 1959, 55°, 0".4.
STF 2843	ADS 15407. 1831, 134°, 2".4.
STT 458	ADS 15481. Fixed. Comes 12.2 at 23".
BU 275	ADS 15499. 1876, 183°, 0".3.

PAIR	NOTES
STF 2863	ADS 15600. Xi Cep. P=3800 years? w,b. Comes 12.7 at 97".
STF 2881	ADS 15769. 1830, 111°, 1".8.
HO 180	ADS 15794. 1886, 222°, 0".5.
BU 1216	ADS 15843. 1890, 318°, 0".6.
BU 172	ADS 15902. 51 Aqr. 1875, 20°, 0".5. Long period binary. Comites 10.1 at 54", 10.0 at 116" and 8.5 at 132".
KR 60	ADS 15972. P=44.5 years. (1975.0 = 1".84). Fainter star is DO Cep, a flare star. Suspected dark comes to A - many other comites in the IDS.
STF 2909	ADS 15971. Zeta Aqr. P=856 years? Perturbation of B with a period of 25.5 years. ysh,bsh.
STF 2912	ADS 15988. 37 Peg. P=140 years. (1975.0 = 1".07). Orbit of high inclination.
STF 2920	ADS 16069. Fixed. Py,Pb.
HO 296	ADS 16173. P=20.93 years. (1975.0 = 0".47). y,y. Third body suspected from period variation.
STF 2934	ADS 16185. P=520 years. (1975.0 = 0".87). A has a dark companion with a period of 81 years.
STF 2943	ADS 16268. Tau^1 Aqr. 1783, 110°, 35".6.
STF 2950	ADS 16317. 1832, 319°, 2".0. y,b. Comes 10.7 at 39".
A 632	ADS 16326. P=90 years or 98.3 years.
COU 240	ADS - . Discovered in 1968.
STT 483	ADS 16428. 52 Peg. P=286 years. (1975.0 = 0".68).
HLD 56	ADS 16435. 1881, 125°, 0".9.
STF 2974	ADS 16496. Fixed.
BU 385	ADS 16561. 1876, 136°, 0".4. Comes 9.0 at 58" forms HJ 5532.
STF 2993	ADS 16611. AB fixed. Py,p. AC, 1824, 109°, 158" forms S 826 - distance decreasing due to pm of AB.
STF 3001	ADS 16666. Omicron Cep. P=796 years. (1975.0 = 2".91). Py,Pb. Comes 12.8 at 46".
BU 80	ADS 16665. P=91.8 years. (1975.0 = 0".97). Eccentric orbit. Comites 10.0 at 106", 9.4 at 210", 9.8 at 206".
STF 2998	ADS 16672. 94 Aqr. 1830, 345°, 13".4. Py,Pb.
STT 494	ADS 16686. Fixed.
STF 3007	ADS 16713. 1829, 79°, 5".7. Comes 9.0 at 91" (dist. inc.) Main pair y,b.
STT 496	ADS 16795. 1 Cas. 1851, 337°, 1",5. (AB). Multiple star - six other components.
BU 720	ADS 16836. 72 Peg. P=199 years or 425 years. y,y.
STT 500	ADS 16877. 1845, 299°, 0".4. g,g. Comes 10.5 at 116".
BU 858	ADS 16928. 1881, 277°, 0".5. Comes 12.3 at 23" forms BU 389.
STT 510	ADS 17050. 1848, 348°, 0".4. Comes 9.0 at 21" forms HJ 1911.
STF 3050	ADS 17149. P=320 years. bw,bw. Comes 12.9 at 81".

APPENDIX I.

References & Bibliography.

Chapter 1. (Historical)

Aitken, R.G.	The Binary Stars.	Dover.	1963.
Crossley, E., Gledhill, J., Wilson, J.H.	A Handbook of Double Stars.	Macmillan.	1879.
Dembowski, E.	Misure Micrometriche di Stelle Doppie e Multiple.	Rome.	1883.
Herschel, W.	Phil. Trans. Vol.72. p112.		1782.
	p339.		1803.
Mackie, J.	BAAJ, Vol.80. p48.		1969.

Chapter 2. (Types of Double Stars)

Evans, D.S.	Observations in Modern Astronomy. p189.	English University Press.	1968.

Chapter 3. (Observing Method)

Aitken, R.G.	The Binary Stars.	Dover.	1963.
Hartung, E.J.	Astronomical Objects for Southern Telescopes.	Cambridge University Press.	1968.
Lewis, T.	Dawes's Limit for Unequal Pairs. Observatory, Vol.37. p378.		1914.
Treanor, P.J.	On the Telescopic Resolution of Unequal Binaries. Observatory, Vol.66. p255.		1946.

Chapter 5. (Micrometers for Double Star Measurement)

Diffraction Micrometer.

Schwarzchild, K.	Astr. Nach. Vol.139. p353.		1896.
Richardson, L.	BAAJ, Vol.35. p105.		1924.
	Vol.35. p155.		1925.
	Vol.37. p311.		1927.
	Vol.38. p258.		1928.
Duruy, M.V.	Private correspondence.		1972.

References & Bibliography.

Chapter 5. (Micrometers for Double Star Measurement) Continued.

Binocular Micrometer.

Duruy, M.V.	J. des Observateurs. Vol.21. p97.	1938.
Duruy. M.V.	WSQJ, Vol.5. No 2.	1972.
Duruy, M.V.	Private correspondence.	1972.

Filar Micrometer.

Aitken, R.G.	The Binary Stars. Dover.	1963.
Jonckheere, R.	J. des Observateurs. Vol.33. p57.	1950.
Muller, P.	Stars. Stellar Systems. Vol.3. Ch19. p440. Un. of Chicago Press.	
Worley, C.E.	Sky & Telescope. Aug 1961. p 73. Sep 1961. p140. Nov 1961. p261.	

Double Image Micrometer.

Airy, G.B.	Memoirs RAS, Vol.15. p199.	1846.
Muller, P.	Stars. Stellar Systems. Vol.3. Ch19. p446. Un. of Chicago Press.	
Muller, P.	Comptes Rendus. Vol.205. p961.	1937.

Comparison Image Micrometer.

Hargreaves, F.	MNRAS. Vol.92. p 72.	1931.
	ibid. p453.	1932.
Carmichel, H.	J. des Observateurs. Vol.39. p138.	1956.
Davidson, C.R., Symms, L.S.T.	MNRAS. Vol.98. p176.	1937.

Interferometers.

Anderson, J.A.	Ap. J. Vol.51. p263.	1920.
Finsen, W.S.	MNRAS. Vol.111. p391.	1951.
Finsen, W.S.	Popular Astronomy, Vol.59. p399.	1951.
Gezari, D.Y., Labeyne, A., Stachnik, R.V.	Ap. J. Letters. Vol.173. No 1. Part 2. Page L1.	1972.

References & Bibliography.

Muller, P.	Stars. Stellar Systems. Vol.3. CH19. p452.	Un. of Chicago Press.	
Michelson, A.	Ap. J. Vol.51. p257.		1920.

Chapter 6. (Double Star Photography)

Sidgwick, J.B.	Amateur Astronomers Handbook.	Faber & Faber.	1971.
Rackham, T.W.	Astronomical Photography at the Telescope.	Faber & Faber.	1959.
Merton, G.	BAAJ, Vol.63. p7.		1952.

Chapter 8. (Double Star Catalogue)

Batten, A.H.	Binary and Multiple Systems of Stars.	Pergamon.	1973.
Binnendijk, A.	The Properties of Double Stars.	University of Pennsylvania Press.	1960.
Candy, M.P.	Practical Amateur Astronomy. p175.	Lutterworth Press.	1963.
Finsen, W.S., Worley, C.E.	A Third Catalogue of Visual Binary Orbits. Republic Obs. Circ. Vol.7. No 129.		1970.
Hartung, E.J.	Astronomical Objects for Southern Telescopes.	Cambridge University Press.	1968.
Hussey, W.J.	Micrometrical Measures of the Double Stars discovered at Poulkova. Publ. Lick Obs. Vol.5.		1901.
Jeffers, H.M., van den Bos, W.H. Greenby, F.M.	Index Catalogue of Double Stars. 1961.0. Publ. Lick Obs. Vol.21.		1963.
Lewis, T.	Measures of the Double Stars contained in the Mensurae Micrometricae of F.G.W. Struve. Publ. RAS. Vol.56.		1906.
Meeus, J.	Some Bright Visual Binary Stars. Sky & Telescope. Jan 1971. p21. ibid. Feb 1971. p88.		

References & Bibliography.

Chapter 8. (Double Star Catalogue) Continued.

Muller, P., Meyer, C.	Troisièmè Catalogues D'Ephemerides D'Etoiles Doubles. Publ. Obs. of Paris.		1969.
Sidgwick, J.B.	Amateur Astronomers Handbook.	Faber & Faber.	1971.
Sidgwick, J.B.	Observational Astronomy for Amateurs.	Faber & Faber.	
Worley, C.E.	Sky & Telescope. Aug 1961. p73.		